Boronic Acids in Saccharide Recognition

Monographs in Supramolecular Chemistry

Series Editor
J Fraser Stoddard, FRS, *University of California at Los Angeles, USA*

This series has been designed to reveal the challenges, rewards, fascination and excitement in this new branch of molecular science to a wide audience and to popularize it among the scientific community at large.

Titles in this series:

Anion Receptor Chemistry
Jonathan L. Sessler, *University of Texas, Austin, Texas, USA*, Philip A. Gale, *University of Southampton, Southampton, UK and* Won-Seob Cho, *University of Texas, Austin, Texas, USA*
Boronic Acids in Saccharide Recognition
Tony D. James, *Department of Chemistry, University of Bath, Bath, UK*, Marcus D. Phillips, *Department of Chemistry, University of Bath, Bath, UK and* Seiji Shinkai, *Department of Chemistry and Biochemistry, Graduate School of Engineering, Kyushu University, Fukuoka, Japan*
Calixarenes
C. David Gutsche, *Washington University, St Louis, USA*
Calixarenes Revisited
C. David Gutsche, *Texas Christian University, Fort Worth, USA*
Container Molecules and Their Guests
Donald J. Cram and Jane M. Cram, *University of California at Los Angeles, USA*
Crown Ethers and Cryptands
George W. Gokel, *University of Miami, USA*
Cyclophanes
François Diederich, *University of California at Los Angeles, USA*
Membranes and Molecular Assemblies: The Synkinetic Approach
Jürgen-Hinrich Fuhrhop and Jürgen Köning, *Freie Universität Berlin, Germany*
Self-Assembly in Supramolecular Systems
Len Lindoy, *The University of Sydney, Australia* and Ian Atkinson, *James Cook University, Townsville, Australia*

Visit our website at www.rsc.org/Publishing/Books/MOSC

How to obtain future titles on publication
A standing order plan is available for this series. A standing order will bring delivery of each new volume immediately on publication.

For further information please contact:
Sales and Customer Care, Royal Society of Chemistry, Thomas Graham House
Science Park, Milton Road, Cambridge, CB4 0WF, UK
Telephone: +44 (0)1223 432360, Fax: +44 (0)1223 426017, Email: sales@rsc.org

Boronic Acids in Saccharide Recognition

Tony D. James
Department of Chemistry, University of Bath, Bath, UK

Marcus D. Phillips
Department of Chemistry, University of Bath, Bath, UK

Seiji Shinkai
Department of Chemistry and Biochemistry, Graduate School of Engineering, Kyushu University, Fukuoka, Japan

RSCPublishing

ISBN-10: 0-85404-537-6
ISBN-13: 978-0-85404-537-2

A catalogue record for this book is available from the British Library

Published by The Royal Society of Chemistry,
Thomas Graham House, Science Park, Milton Road,
Cambridge CB4 0WF, UK

Registered Charity Number 207890

For further information see our web site at www.rsc.org

Typeset by Macmillan India Ltd, Bangalore, India
Printed by Henry Ling Ltd, Dorchester, Dorset, UK

Preface

The ability to monitor the presence of analytes within physiological, environmental and industrial systems is of crucial importance. However, owing to the scale that recognition events occur at on the molecular level, gathering this information poses a non-trivial challenge. It is therefore the case that robust chemical molecular sensors with the capacity to detect chosen molecules selectively and signal this presence have attracted considerable attention over recent years.

Of particular interest is the real-time monitoring of saccharides in aqueous systems, such as D-glucose in blood. To this end, the covalent pair-wise interaction between boronic acids and saccharides has been exploited.

This book documents research into the design of novel boronic acid-based receptors with selectivity for saccharides.

There is so much to do. You can wander off in space or in time, set out for Tierra del Fuego or for King Midas's court . . . You can build castles in Spain, steal the Golden Fleece, discover Atlantis, realise your childhood dreams and adult ambitions.

Jean-Dominique Bauby
The Diving-bell and the Butterfly

Contents

CHAPTER 1

Introduction

What is now proved was once only imagined

William Blake, 1757–1827

The recognition of a target molecule by a synthetically prepared receptor has captured the imagination of supramolecular chemists. Since its inception, research in this area has been instrumental in elucidating the mode of action of a great many biological events concerning recognition and catalysis.[1] The importance of this work was underlined with the award of the Nobel Prize in Chemistry to Cram, Lehn and Peterson in 1987 "for their development and use of molecules with structure-specific interactions of high selectivity."[2] Since then the diversity of compounds studied under the umbrella of supramolecular chemistry has grown significantly. Of particular interest are chemical molecular sensors, single molecules with the ability to both recognise and signal analytes in real time.[3,4]

The development of coherent strategies for the selective binding of target molecules, by rationally designed synthetic receptors, remains one of chemistry's most sought after goals. The research conducted to this end is driven by a fundamental inquisitiveness and need to monitor compounds of industrial, environmental and biological significance.

Within our research groups we have exploited the interaction between boronic acids and diols. The primary interaction of a boronic acid with a diol is covalent and involves the rapid and reversible formation of a cyclic boronate ester. The array of hydroxyl groups present on saccharides provides an ideal scaffold for these interactions and has led to the development of boronic acid-based sensors for saccharides (Scheme 1).[5–18]

Many synthetic receptors developed for neutral guests have relied on non-covalent interactions, such as hydrogen bonding, for recognition. It is the case, however, that in aqueous systems neutral guests may become heavily solvated. While biological systems have the capacity to expel water from their binding pockets and sequester analytes wholly, using non-covalent interactions, synthetic monomeric receptors have not yet been designed where hydrogen

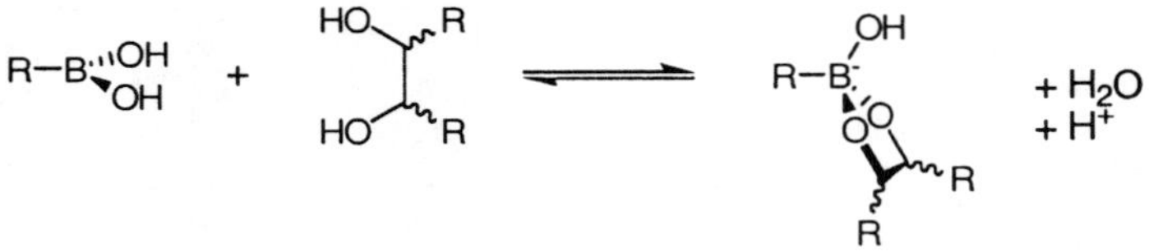

Scheme 1 *The rapid and reversible formation of a cyclic boronate ester.*

bonding has been able to compete with bulk water for low concentrations of monosaccharides.[19] However, it should be pointed out that progress is being made in this area and recently Davis reported a hydrogen-bonding receptor capable of binding D-glucose in water with a weak but significant stability constant.[20]

The capacity of boronic acid receptors to function effectively in water is reflected by the number of published sensory systems designed around them. The most popular class of the fluorescent boronic acid-based sensors utilise an amine group proximal to boron. The Lewis acid–Lewis base interaction between the boronic acid and the neighbouring tertiary amine has a dual role. First, it enables molecular recognition to occur at neutral pH. Second, it can be used to communicate binding by modulating the intensity of fluorescence emission through photoinduced electron transfer (PET), introducing an "off–on" optical response to the sensor.

The quality of the research in this area, particularly in the past few years, has led to a significant advance in the understanding of the basic science behind the generic mode of action of this class of receptor. *This book will therefore bring together and critique the contemporary scientific understanding of the fundamental processes involved in the molecular recognition of saccharides by boronic acids.* It should be noted that a comprehensive overview of this nature has not been previously reported in the scientific literature. *A literature review will then investigate the application of these sensory systems.* Although this review cannot be exhaustive, it is our intention to illustrate the current state of play in the field.

CHAPTER 2

The Molecular Recognition of Saccharides

The expression "fits like a glove" is an odd one, because there are many different types of gloves and only a few of them are going to fit the situation you are in.

Lemony Snicket
A Series of Unfortunate Events: The Grim Grotto: Book the Eleventh

2.1 Molecular Recognition

Molecular recognition lies at the very heart of sensor chemistry. The process itself involves the interaction between two substances, often termed as a host and a guest, a lock and a key or a receptor and a substrate. Importantly, recognition is not just defined as a binding event but requires selectivity between the host and the guest.

Selectivity between host and guest is a premise of compatibility. It arises between compounds with carefully matched electronic, geometric and polar elements. For synthetic receptors, the potential, therefore, exists to engineer receptors for any chosen analyte through judicious structural design and functional group complementarity. The power of this concept is illustrated within Nature. Biological systems have evolved with exquisitely constructed active-binding sites, sequestering guest molecules with near perfect selectivity.

Nevertheless, for the recognition event at a receptor to be of practical use, a further element is required. A channel of communication must be established between the receptor and the outside world. This additional quality converts a receptor into a sensor.

For a sensor to function, it must, therefore, permit selective binding to occur between host and guest and also report these binding events by generating a tangible signal. By performing these two fundamental tasks, sensors have the potential to relay information on the presence and location of important species in a quantifiable manner, bridging the gap between events occurring at the molecular level and our own (see Scheme 2).

Chemical sensors can be broadly categorised as either biosensors, or synthetic sensors. Biosensors make use of existing biological elements for recognition. Many of the physiologically important analytes already have corresponding biological receptors with intrinsically high selectivity and if

3

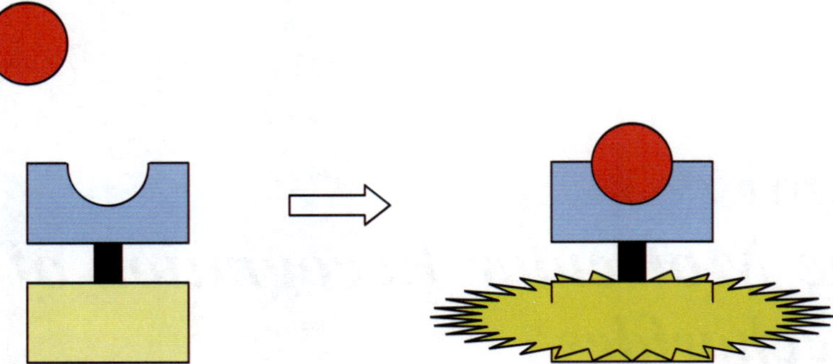

Scheme 2 *The complementary interaction between a guest analyte and a host-binding pocket, illustrated here by a red guest and blue host, allows selective binding to occur between two elements. Incorporation of a unit capable of generating a physical signal in response to this binding event, converts the receptor into a sensor. In this cartoon, an optical "off-on" response is depicted from an appended fluorophore, illustrated in yellow, typical of the systems covered within this book.*

these receptors can be connected to a signal transducer, a biosensor can be developed.[21]

Synthetic sensors incorporate a synthetically prepared element for recognition. While biomimetic receptors have been prepared, with synthetic receptors mimicking the active sites in naturally occurring biological molecules, synthetic receptors can, and often are, designed entirely from first principles.

2.2   The Importance of Saccharides

2.2.1   Saccharides and Carbohydrates

In keeping with convention, the term saccharide is used within this book to refer broadly to polyhydroxylated carbohydrates.[22] The product of photosynthesis, carbohydrates single-handedly account for the most prolific class of organic compounds that can be found on the surface of the Earth. Within biology they are of fundamental significance. In their most ubiquitous roles, they endow Nature with structural rigidity, in the form of cellulose, and function as the energy store that sustains life, in the forms of starch and glycogen.[23]

Not only are these compounds abundant they are also incredibly versatile. Oligosaccharides are involved in protein targeting and folding, as well as controlling the cell recognition events for infection, inflammation and immunity.[24] From a medicinal perspective, the monitoring of D-glucose has proved of particular importance. D-glucose provides the metabolic energy for most cells of higher organisms. In humans, a breakdown in the transport pathways of D-glucose has been linked to conditions such as cancer,[25] cystic fibrosis[26] and renal glycosuria,[27,28] but by far the most prevalent condition resulting from ineffective D-glucose transport is diabetes mellitus.[29]

2.2.2 Diabetes Mellitus

Diabetes presents one of the largest health challenges to face us in the 21st century. Current reports indicate that diabetes affects 5% of the global population.[30] In the UK, the increase in obesity, population age and a progressively more sedentary lifestyle has seen the prevalence of Type 1 diabetes double every 20 years since 1945.[31] Diabetes is associated with chronic ill health, disability and premature mortality. From a physiological perspective, the debilitating long-term complications include heart disease,[32] blindness,[33] kidney failure,[34] stroke[35] and nerve damage leading to amputation.[36]

At an economic level the repercussions are also serious. Within the UK, 5% of the National Health Service's budget is spent on treating diabetes and its complications.[37] This equates to £3.5 billion per year or £9.6 million per day. Following extensive and widespread trials, unequivocal evidence exists that monitoring and adjusting diabetic blood sugar levels to maintain them within tight boundaries[†] dramatically reduces the health risks faced by diabetics.[39–41]

2.2.3 Structure of Saccharides

The detection of saccharides presents a curious challenge. Bristling with linked arrays of hydroxyl groups, saccharides are structurally complex. The linear form of D-glucose, for example, contains four stereocentres. Considering just its immediate family, the aldohexoses, this presents us with 16 stereoisomers (see Figure 1).

In aqueous solution, this complexity is further compounded by mutorotation. Cleavage of the hemiacetal ring causes interconversion between the pyranose and furanose ring forms, *via* an acyclic intermediate, with inversion of configuration at the anomeric centre equilibrating the α- and β-anomers thus altering the optical rotation of the solution (see Figure 2). These properties make saccharides difficult to differentiate from each other. Furthermore, in water receptors for saccharides may be heavily solvated, differentiation between water and hydroxyl groups presenting a formidable challenge in its own right.

This point can be illustrated by considering the protein family, lectins. Aside from antibodies, lectins are largely responsible for carbohydrate recognition within biological systems. Lectin–oligosaccharide-binding constants are, by biological standards, unusually small, commonly in the range 10^3–10^4 M^{-1}.

The crystal structure of the *Escherichia coli* galactose-binding protein with a bound molecule of β-D-glucose was obtained.[44] From the structure, it can be observed that recognition is achieved by sequestering the monosaccharide entirely beneath the protein surface and expelling bulk water from the active site. This not only fosters a favourable entropic stabilisation but also allows the surrounding amino acid residues unfettered access to the substrate. In the

---

[†] non-diabetic blood glucose concentrations are usually in the range 4–7 mM.[38]

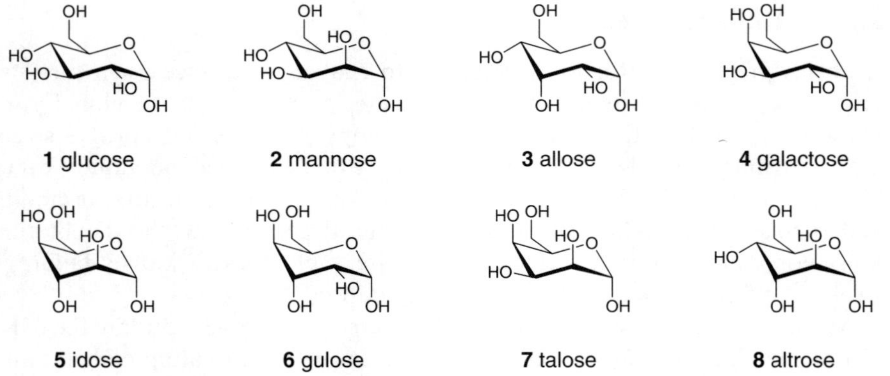

1 glucose   **2** mannose   **3** allose   **4** galactose

5 idose   **6** gulose   **7** talose   **8** altrose

Figure 1 *The α-D-pyranose forms of the aldohexoses in their $^4C_1$ conformations.[42] Glucose is displayed on the top row with (from left to right) its 2-, 3- and 4-epimers. Idose is displayed on the bottom row with (from left to right) its 2-, 3- and 4-epimers. From the structures, the similarity between these monosaccharides, their affinity for water and the complexity derived from their numerous stereogenic centres is apparent.*

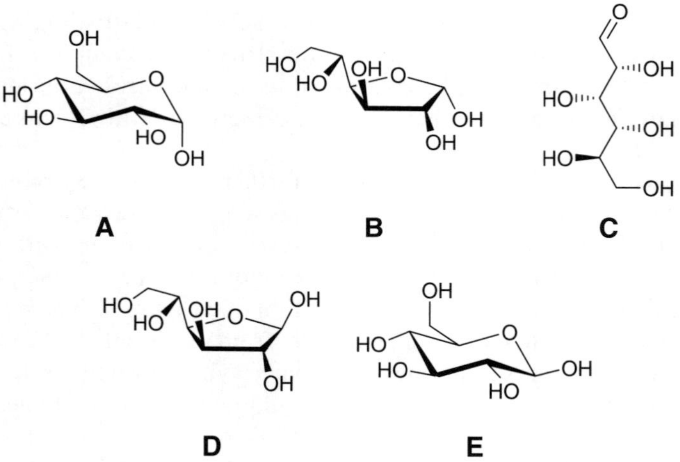

A   **B**   **C**

D   **E**

Figure 2 *D-Glucose* **1** *in its various configurations and the percentage composition at equilibrium of each form of the sugar in D_2O at 27°C: (A) α-pyranose, 38.8%; (B) α-furanose, 0.14%; (C) acyclic form, 0.0024% (equilibrated at 37°C); (D) β-furanose, 0.15%; (E) β-pyranose, 60.9%.[43]*

binding site of the molecule β-D-glucose is sandwiched between two apolar phenylalanine and tryptophan residues axial to the ring. Equatorially, an array of 13 different complementary hydrogen-bonding interactions is provided by one advantageously sited water molecule and eight individual amino acid residues, all ideally located (see Figure 3).

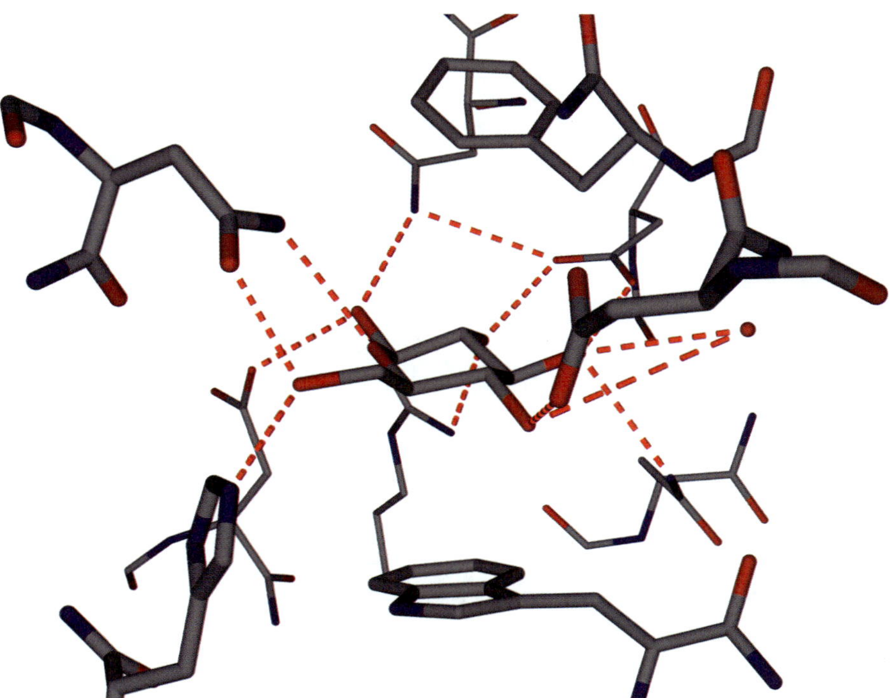

Figure 3 *Crystal structure of the E. coli glucose/galactose-binding protein with β-D-glucose in situ.[44] For clarity, all of the amino acids bar those involved in direct interactions with the saccharide have been stripped away. Atoms marked in red represent oxygen, blue nitrogen and grey carbon, hydrogen atoms are not displayed. The dashed red lines represent hydrogen bonds. Within the active site, the monosaccharide is intercalated between two apolar residues. Axially, below the guest, the indole ring of tryptophan is visible, with the phenyl ring of phenylalanine above. Equatorially, an array of 13 hydrogen bonds provides a cooperative network from carbonyl groups, primary amides, a guanidinium, an imidazole amine and one favourably coordinated water molecule. The allosteric approach used within this active site displays the degree of sophistication and elegance Nature uses to achieve binding.*

2.2.4 Home Blood Glucose Monitoring

Commercially, the preferred tools for sensing complex molecular species have relied on the high specificity exhibited by antibodies and enzymes. Most clinical systems currently available for measuring blood glucose levels rely on the glucose oxidase enzyme (GOx, also commonly abbreviated to GOD or GO).

The majority of these home blood glucose monitoring tools rely on the invasive withdrawal of blood, typically from a pricked finger, followed by application of the sample to an amperometric enzymatic test strip allowing GOx to catalyse the oxidation of glucose to gluconic acid (see Scheme 3).[38,45]

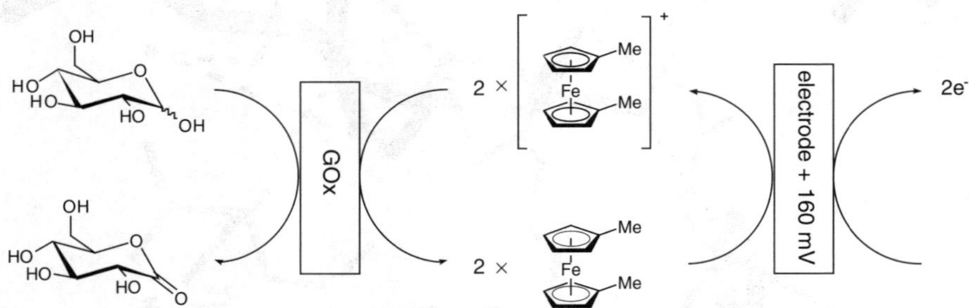

Scheme 3 *The oxidation of D-glucose to D-gluconolactone (the ring closed form of D-gluconic acid).*

Scheme 4 *The oxidation of glucose with concurrent reduction of the dimethylferricinium cation mediator.*

Early glucose monitors measured the production of hydrogen peroxide by oxidation at a single-working electrode, as in Equation (1). At a constant voltage, the current generated across the cell is proportional to the concentration of hydrogen peroxide, which is in turn proportional to the glucose concentration in the sample under investigation.

$$H_2O_2 \rightarrow O_2 + 2H^+ + 2e^- \tag{1}$$

As this method of monitoring glucose is heavily dependent on the oxygen concentration, a dimethylferrocene mediator was developed. At a set potential of +160 mV glucose is oxidised with concurrent reduction of the dimethylferricinium cation; the applied potential serving to oxidise and thus recycle the resulting dimethylferrocene back to the dimethylferricinium cation (see Scheme 4).[46,47]

There can be no question that the availability of affordable home blood glucose monitoring has revolutionised the quality of life experienced by diabetics. However, there are some inherent limitations with an enzymatic approach. The systems have to be stored appropriately, they are specific only for a few saccharides and in most cases they become unstable under harsh conditions and hence cannot be sterilised. For this reason, much work has been focused on the development of synthetic sensors with the capacity to monitor saccharides under a broad range of environmental conditions, and thus allow access to a wider spread of diagnostic applications.

2.3 Non-Boronic Acid Appended Synthetic Sensors for Saccharides

A great amount of attention is currently devoted to the development of synthetic molecular receptors with the ability to recognise saccharides. This book is primarily concerned with the role of boronic acids in that recognition process, although many systems have been developed which use non-covalent interactions for recognition. A few examples are given below to provide a brief illustration of research in this area. (For a more comprehensive insight, the reader is directed to two recently published reviews.[9,19])

The first synthetic saccharide receptor **9** was documented in 1988 by Aoyama *et al.*[48] The calixarene framework is not large enough to actually permit the encapsulation of a saccharide within the central annulus of the bowl, but it does permit "face-to-face" recognition to occur between the calixarene's upper rim hydroxyls and those of the saccharide. However, in order to complex saccharides in water Aoyama needed to use very high concentrations of saccharides (100 mM^{-1} M).

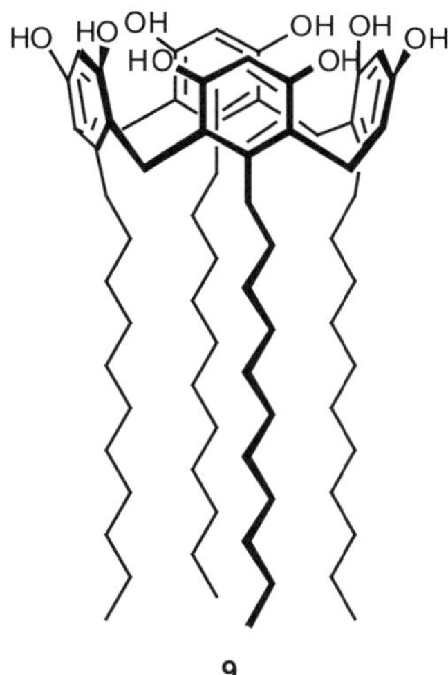

9

Inspired by the structure of the saccharide-binding proteins such as the active site of the *Escherichia coli* galactose-binding protein, illustrated in Figure 3, Davis and Wareham developed the octaamide **10**.[49] The sophisticated architecture mimics Nature in that the binding pocket is designed so as to completely encapsulate a host saccharide. The planar hydrophobic surfaces of the dual diphenyl groups provide internal apolar contacts above and below the aliphatic ring. Located equatorially to the saccharide host an array of secondary amides induces favourable hydrogen-bonding interactions, anchoring the hydroxyls on the saccharide ring.

10: X = OC$_5$H$_{11}$

11

X= **12**

Even with such elegant design the reported stability constants (K_a) for sensor **10** in deuterated chloroform are reduced significantly on addition of competitive co-solvents. In the presence of octyl-β-D-glucoside **11** $\sim$ 300-fold reduction was witnessed in the stability constant (K_a) of sensor **10** on addition of 8% deuterated methanol to the system.[‡] The stability constants (K_a) for sensor **10** with octyl-β-D-glucopyranoside **11** were 300,000 M^{-1} in deuterated chloroform and $\sim$ 1000 M^{-1} in 92:8 deuterated chloroform/deuterated methanol. More recently, the dodecacarboxylate version of this receptor **12** has been prepared and this system does bind D-glucose in water with a stability constant (K_a) of 9.5 M^{-1}.[20] This is much better than any of the previous systems but the binding is still to weak in water for these systems to be useful in detecting saccharides in biological systems.

As well as developing recognition sites through careful structural pre-organisation, non-covalent interactions of increased strength have also been studied. Manipulation of anionic centres has proved successful as functional groups such as phosphates, phosphonates and carboxylates can be powerful hydrogen-bond acceptors. While the pre-organisation can be reduced, it has been found that these anionic centres provide a degree of structure in their own right. Flexible sensors with multiple anionic groups will conform to the tendency of the charged groups to repel each other so as to minimise electrostatic repulsions.

Das and Hamilton have employed phosphonate moieties in sensors such as the rigid chiral spirobifluorene **13**.[50] The racemate was used directly and although the exact nature of the hydrogen-bonding motif could not be determined excellent binding was observed with octylglucosides in organic solvents.

[‡] Unless specified, otherwise all of the deuterated solvents referred to in this book are fully deuterated.

The enantioselectivity was also determined and reported as $\sim 5.1{:}1$, the greatest value reported for saccharide enantioselective discrimination by non-covalent interactions.[9] The binding constant (K_a) for sensor **13** with octyl-β-D-glucopyranoside **11** was 47,000 M^{-1} in deuterated acetonitrile.

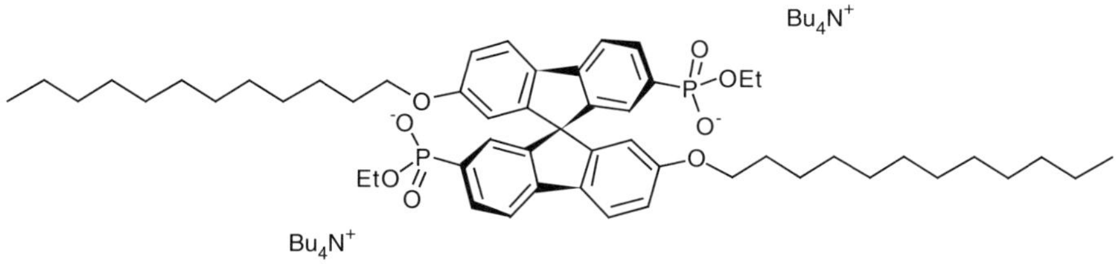

13

Diederich and co-workers have made use of binaphthalene-derived macro-cycles appended with internal phosphate groups to provide a ring of hydrogen-bonding sites within a central recognition cavity.[51–53] One of the first of these, sensors **14**, was shown to have good selectivity for suitably sized saccharides such as the octyl-D-β-glucopyranoside **11**, the calculated interphosphate distance of 7.2 Å being designed so as to accommodate monosaccharides exclusively. A stability constant (K_a) of 5200 M^{-1} was reported for the complexation of sensor **14** and octyl-D-β-glucopyranoside **11** in 98:2 deuterated acetonitrile/deuterated methanol.[53]

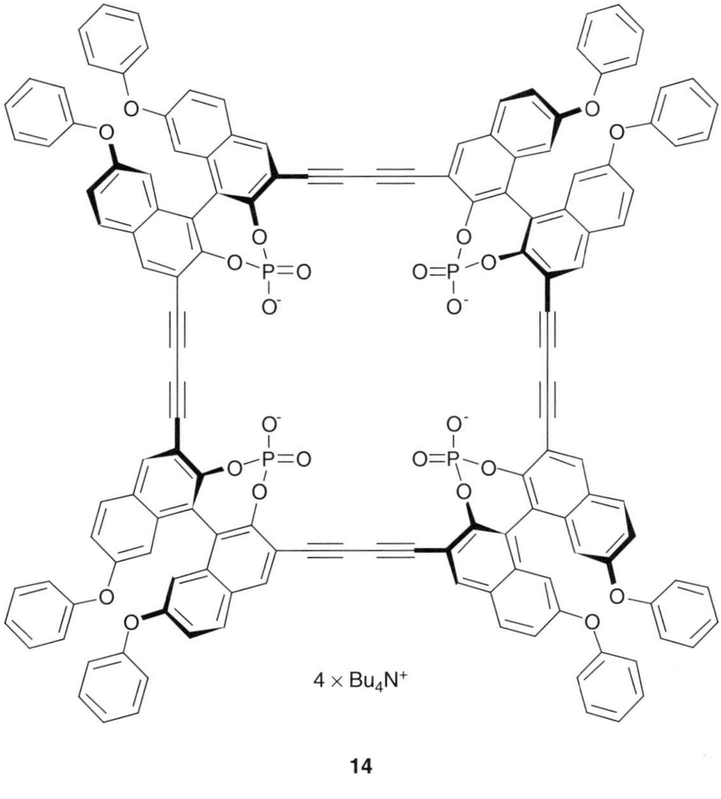

14

The major obstacle faced by all of the synthetic receptors above, which are wholly reliant on non-covalent interactions, is that of solvent competition. In aqueous systems, neutral guests may become heavily solvated and are therefore unable to monitor glucose in media of specific interest to analysts such as blood, urine, tear fluid, beverages, foodstuffs and so forth. Because, monitoring saccharide interactions in water with these receptors is difficult they are typically evaluated in aprotic solvens such as chloroform. By using aprotic solvents such as chloroform simple saccharides can not be used because they are insoluble, when using these solvents it is necessary to use soluble O-alkylated saccharides.[9,19] In overcoming this hurdle, the covalent interaction between boronic acids and saccharides has proved hugely advantageous.

CHAPTER 3

Complexation of Boronic Acids with Saccharides

In the field of observation, chance favours only the prepared mind
Louis Pasteur, 1822—1895

3.1   A Brief History

3.1.1   Early Work

Boronic acid chemistry has its roots in the work of Frankland, who in 1860 documented the preparation of ethylboronic acid, with the first synthetic publication on organoboron chemistry (The structures of boric acid and phenyl boronic acid are shown in Figure 4).[54] In 1880 Michaelis and Becker reacted borontrichloride and diphenyl mercury to form dichlorophenylborane. This in turn was added to water and recrystallised as white needles in the first synthesis of phenylboronic acid.[55,56] The route was refined in 1909 and the classical synthesis of boronic acids from Grignard reagents and trialkyl borates was established.[57]

In a pioneering series of papers spanning 29 years, from 1911 to 1940, Böeseken and co-workers elucidated the absolute structural configuration of a panoply of saccharides and other hydroxyl-containing compounds.[58] The configurations were determined, in large part, from the known ability of boric acid to complex diols. As boric acid complexes formed in solution, with hydroxyl groups in favourable 1,2- and sometimes 1,3- configurations, a rise was observed in both acidity and conductivity. By tracking the changes in conductivity of boric acid solutions as saccharides were added, the relative configurations of the hydroxyl groups could be inferred. This method was employed in 1913 to conclusively determine the structure of the pyranose and furanose forms of D-glucose. Changes in conductivity revealed the relative *cis*- and *trans*-orientations of the vicinal C1 and C2 hydroxyl groups, thus

15   16

Figure 4   *The planar structures of boric acid* **15** *and phenylboronic acid* **16**.

13

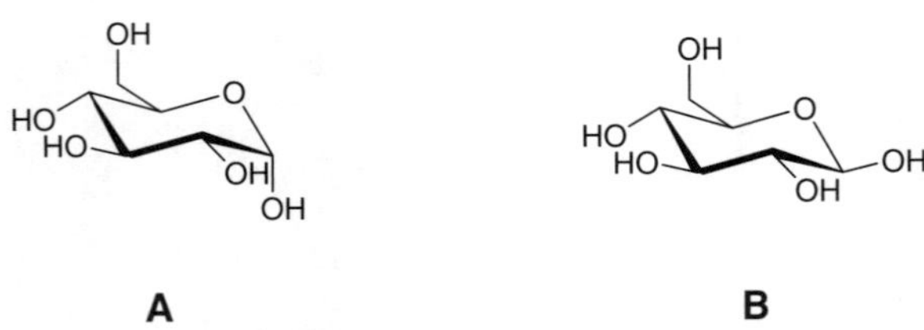

A   **B**

Figure 5   *The pyranose anomers of D-glucose 1: (**A**) α-D-glucopyranose and (**B**) β-D-glucopyranose. The relative cis- and trans-orientations of the vicinal C1 and C2 hydroxyl groups were inferred by tracking the changes in conductivity of boric acid solutions as these saccharides were added, permitting the absolute structural configuration of the α- and β-anomers to be attributed.[59]*

permitting the absolute structural configuration of the α- and β-pyranose and furanose anomers to be attributed (Figure 5).[59]

3.1.2   Boronic Acid – Diol Complexation

Given the significance of boric acid in the determination of saccharide configurations, it is perhaps surprising that the same properties were not observed in boronic acids until 1954.[60] During a course of investigations into aromatic boronic acids, Kuivila and co-workers observed a new compound being formed on addition of phenylboronic acid to a solution of saturated mannitol, correctly postulating the formation of a cyclic boronic ester analogous to the one known to form between boric acid and polyols.

A number of publications followed examining the properties and synthesis of boronic acids,[61–63] with the first quantitative investigation into the interactions between boronic acids and polyols in 1959.[64] In a study to clarify the disputed structure of the phenylboronate anion, Lorand and Edwards added a range of polyols to solutions of phenylboronic acid. The pH of the solutions was adjusted such that there was an equal speciation of phenylboronic acid in its neutral and anionic forms; the pH matching the pK_a. As diol was added the pH of the systems decreased, allowing binding constants to be determined through the technique of pH depression.

From these experiments, Lorand and Edwards concluded that the conjugate base of phenylboronic acid has a tetrahedral, rather than trigonal structure. The dissociation of a hydrogen ion from phenylboronic acid occurs from the interaction of the boron atom with a molecule of water. As the phenylboronic acid and water react, a hydrated proton is liberated, thereby defining the acidity constant K_a, see Hartley, Phillips and James.[65] This is depicted in Scheme 5 by considering an explicit water molecule associated with the Lewis acidic boron. The reported pK_as of phenylboronic acid fluctuate between ∼8.7 and 8.9,[66–71] with a recent in-depth potentiometric titration study refining this value to 8.70 in water at 25°C.[72]

While one explicitly associated molecule is shown in a number of illustrative schemes for clarity, water should be considered to be in rapid exchange on the

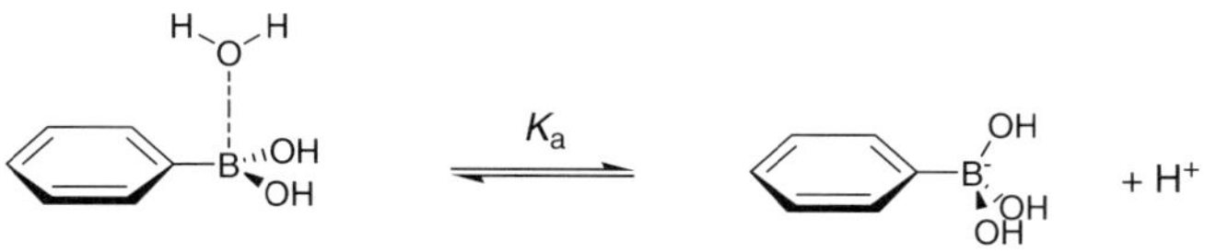

Scheme 5 *The acid–conjugate base equilibrium for phenylboronic acid in water. The dissociation of the hydrogen ion from phenylboronic acid occurs from the interaction of the boron atom with a molecule of water. Here we consider an explicit water molecule associated with the Lewis acidic boron.[65] As the phenylboronic acid and water react a solvated hydrogen ion is liberated, thereby defining the acidity constant Ka, where pKa=8.70 in water at 25°C.[72]*

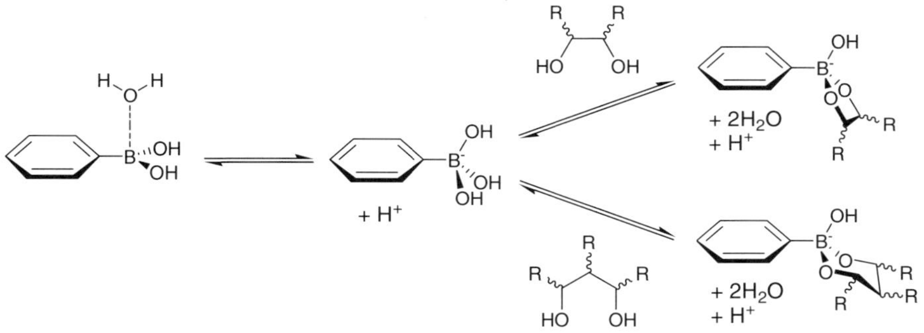

Scheme 6 *Interaction of the phenylboronate anion with 1,2 and 1,3 diols to form diol–phenylboronate complexes with five and six membered rings, respectively. From experimental observations it is well known that the kinetics of this interconversion are fastest in aqueous basic media (by a factor of 10^4) where the boron is present in its tetrahedral anionic form.[80]*

Lewis acidic boron in much the same way that hydrated Lewis acidic metal ions exchange bound water. A pertinent comparison can be found with the ionisation of Zn^{2+} in water, the reaction $Zn–OH_2 \rightarrow Zn–OH + H^+$ having a pK_a of 8.8.[73]

Boronic acids have been reported to rapidly and reversibly[74] interact with dicarboxylic acids,[75,76] α-hydroxy carboxylic acids[70,76–78] and diols[64] to form esters in aqueous media.[79] The most common interaction is with 1,2 and 1,3 diols to form five- and six-membered rings, respectively. From experimental observations, it is well known that the kinetics of this interconversion are fastest in aqueous basic media where the boron is present in its tetrahedral anionic form.[80] Typically, differences in rate of 10^4 are observed between boron in its trigonal and tetrahedral forms.[79]

While the boronate anion does account for the strongest binding of diols in aqueous media, the interaction between diols and the neutral boronic acid should not be ignored. In considering these interactions, the equilibria in Scheme 6 can be readily expanded to form a thermodynamic cycle, as illustrated in Scheme 7.

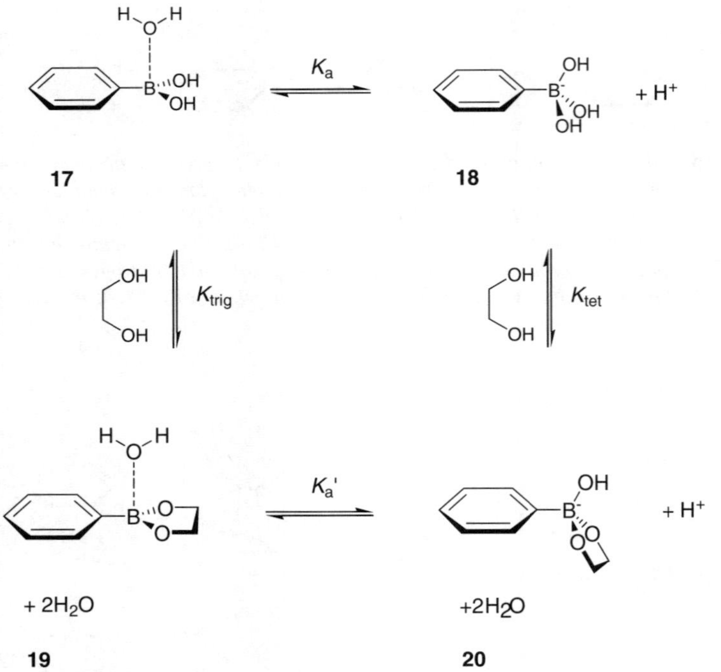

Scheme 7 *The equilibria for boronate ester formation couple to generate a thermody-namic cycle. The formation of the diol boronate anion complex is defined as K_{tet} and the formation of the diol boronic acid complex is defined as K_{trig}, where it is observed that $K_{tet} > K_{trig}$. The acidity constant of the unbound complex is defined as K_a and the acidity constant of the bound complex is defined as K_a', where it is observed that $pK_a > pK_a'$.*

Considering Scheme 7 we define the formation of the diol boronate anion complex as K_{tet} and the formation of the diol boronic acid complex as K_{trig}, where it is observed that $K_{tet} > K_{trig}$. For instance, the logarithm of these constants for phenylboronic acid binding fructose in 0.5 M NaCl water is: log K_{tet}=3.8 whereas log $K_{trig} < -1.4$. This difference in the value of the binding constant between K_{tet} and K_{trig} is typical, with differences of up to approximately five orders of magnitude being commonplace.[72] It is also known that the neutral boronic acid becomes more acidic upon binding. The acidity constant of the bound complex is defined by K_a', where it is observed that $pK_a > pK_a'$. For instance, the pK_a of phenylboronic acid=9.0 in 0.1 M NaCl 1:2 (v/v) methanol/water, under the same conditions the pK_a' of phenylboronic acid bound to fructose=5.2; in other words the boronic ester is more acidic than the boronic acid.[72]

These phenomena were known and employed in analyte recognition for many years prior to an understanding of the mechanistic rationale behind the observed results. Although there are still various points of contention, a great deal of work has been conducted during the last few years, which has helped elucidate the cause of these empirical results. Some of these points are worthy of further scrutiny here.

3.2 Acidity and the O–B–O Bond Angle

3.2.1 O–B–O Bond Angle Contraction

The first is the change in acidity at the boron centre on binding. This change in acidity is often ascribed to the contraction of the oxygen–boron–oxygen (O–B–O) bond angle on complexation and is crucially important to the development of sensory systems containing boronic acid recognition sites as this change in acidity forms the basis of many of the signalling mechanisms used.

In unbound phenylboronic acid the molecule is trigonal planar with sp^2 hybridisation about the boron centre and an expected O–B–O bond angle of $\sim 120°$. Unfortunately, the reported data pertaining to the structure of phenylboronic acid from X-ray crystallography has not proved effective at refining this value. The crystal structure of phenylboronic acid was reported in 1977,[81] the crystallisation process producing cyclic, nearly planar, hydrogen bonded dimers, see Scheme 8.

The phenylboronic acid molecules participate in linear intermolecular hydrogen bonding interactions, a binding motif comparable to the one observed in dimeric carboxylic acids. These interactions distort the O–B–O atoms from ideal geometries (O–B–O bond angles of $\sim 116.3°$ were reported) and void a meaningful comparison between the structure of the crystals examined and the structures that one might expect for the molecules in solution. This dimerisation is commonplace in the crystallisation of arylboronic acids, with similar compressions of the O–B–O bond angles occurring in other compounds.[82] This characteristic has actually found applications in template-directed supramolecular synthesis.[83,84]

In avoiding the structural distortions observed for crystalline phenylboronic acid, the extended crystal lattice formed by boric acid can be considered as an example of a boron species with no deviation from the expected 120° O–B–O bond angle. As hydrogen bonding interactions link all the hydroxyl groups, regular patterns of trigonal and tetrahedral boric acids are fashioned. A recent communication documented the single-crystal X-ray structure of boric acid with C_3 symmetry and an O–B–O bond angle of 120.0 (9)°, as in Figure 6.[85]

On binding, this bond angle is substantially reduced and in doing so places considerable strain on the newly formed ring. Lorand and Edwards estimated

Scheme 8 *Boronic acids exhibit a pre-disposition to associate as cyclic hydrogen bonded dimers upon crystallisation.*

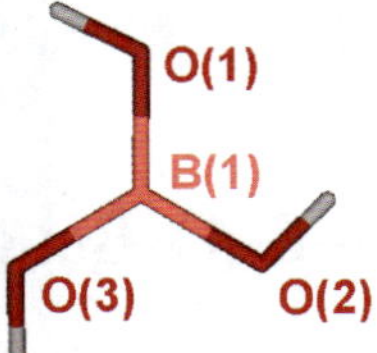

Figure 6 *The single-crystal X-ray structure of boric acid with trigonal symmetry. Atoms marked in red represent oxygen, pink boron and white hydrogen. An O–B–O bond angle of 120.0 (9)° was reported.*

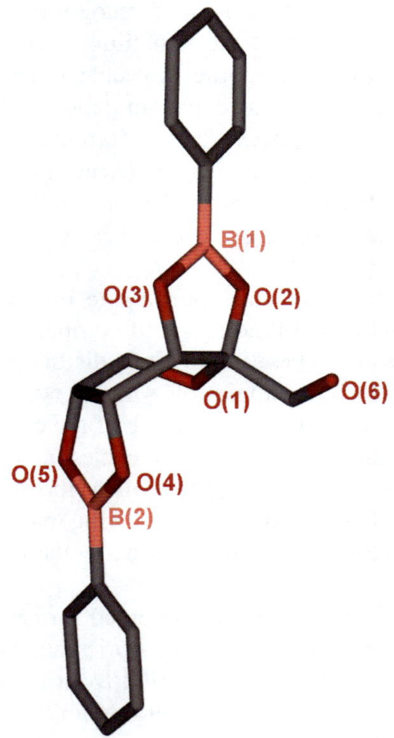

Figure 7 *The single-crystal X-ray structure of the diphenylboronic acid – fructose complex obtained by Fallon and co-workers from a 2 to 1 ratio of phenylboronic acid and D-fructose under non-basic conditions and in the absence of Lewis basic components. Atoms marked in red represent oxygen, pink boron and grey carbon. For clarity hydrogen atoms are not displayed. From the structure it can be observed that the O–B–O bond angles are reduced, in this instance, to ~113° from ~120°, while the structure remains almost entirely planar about the boron centres [O(2)-B(1)-O(3)=113.31(18)° and O(5)-B(2)-O(4)=113.65(18)°].[86]*

that in a five-membered ring an angle of approximately 108° would be obtained. The bond angle is of course heavily dependent on the substrate bound but a good example is the single-crystal X-ray structure of the diphenylboronic acid – fructose complex (Figure 7) obtained by Fallon and

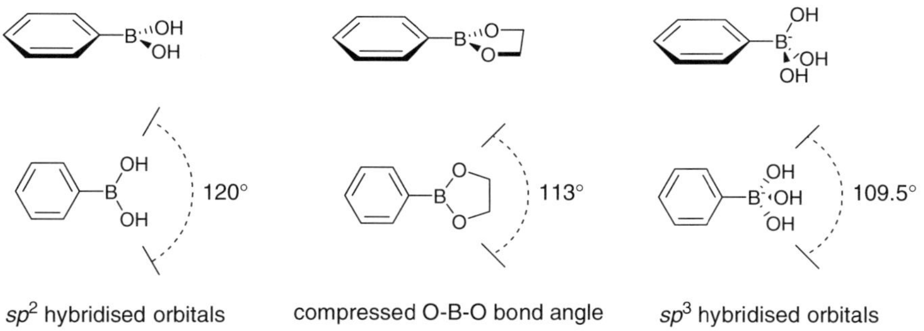

Figure 8 *The compressed O–B–O bond angle of the diphenylboronic acid – D-fructose complex[86] is found to be between the bond angles expected for sp² and sp³ geometry. sp³ Geometry provides the closest match to the bound complex, meaning that formation of the tetrahedral boronate diol complex reduces ring strain and lowers the energy of the bound species. This results in a shift of the dynamic equilibrium between the neutral boronic acid diol complex **19** and the boronate anion diol complex **20**, causing the observed increase in the value of the acidity constant, K_a'.*

co-workers in 2004 from 2 to 1 ratio of boronic acid and D-fructose under non-basic conditions and in the absence of Lewis basic components. From the structure it is observed that five-membered cyclic boronates have formed at the 2,3- and 4,5-positions of β-D-fructopyranose. In the absence of any Lewis basic components, the structure demonstrates little deviation from planarity at the boron centre. Within the plane O–B–O bond angles of just over 113° are observed (Figure 8).[86]

sp^3 Hybridised orbitals are tetrahedral in nature, with an ideal bond angle of 109.5°.[87] The contraction of the O–B–O bond angle means that in the bound instance the geometry of sp^3 hybridised orbitals actually provides a closer topographic match to the complex than the geometry of sp^2 hybridised orbitals.

Rehybridisation of an sp^2 boron complex to form the sp^3 tetrahedral boronate species will therefore reduce the ring strain and lower the energy of the system. As a result it is thought that the dynamic equilibrium illustrated in Scheme 7 (and redrawn in Scheme 9 for clarity) between the neutral boronic acid diol complex **19** and the boronate anion diol complex **20** will shift to the right, causing the observed increase in the value of the acidity constant, K_a'.[8]

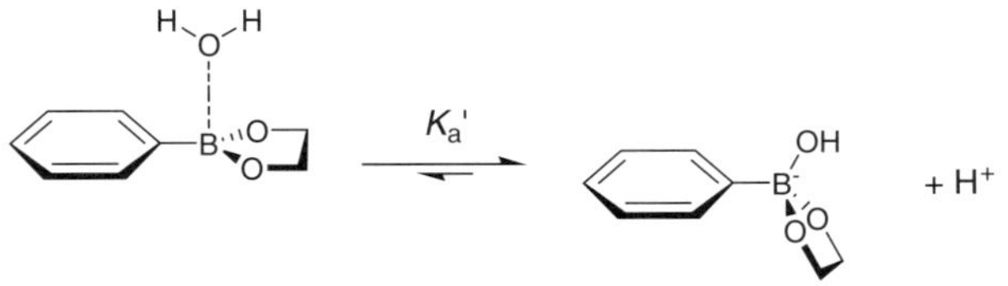

Scheme 9 *The acidity constant of the bound complex is defined as K_a', where it is observed that $pK_a > pK_a'$.*

3.2.2 Orbital Interpretation

The increase in the value of the acidity constant can be predicted somewhat more quantitatively by interpreting the hybridisation of boron's orbitals. In the case of unbound boronic acid, boron has an sp^2 trigonal planar geometry with an empty p orbital perpendicular to the plane of the molecule. It is with this unoccupied p orbital on boron that the nucleophilic oxygen lone pair on the approaching water molecule will mix. As the boron–oxygen interaction strengthens, concomitant proton dissociation occurs. By definition, the ease with which this proton is dissociated determines the value of the acidity constant (Scheme 10).

Following complexation to a diol the geometry of the orbitals at the neutral boron centre are altered. To conform to the constraints enforced upon the boron complex generated, the geometry of the boron orbitals adopt characteristics between sp^2 and sp^3 hybridised orbitals, the empty p orbital therefore develops some s character. It is known that s orbitals are held closer to the nucleus than p orbitals. In the case of filled orbitals this means electrons within s orbitals are lower in energy and more effective at shielding the positive charge of the nucleus than those in p orbitals. In the case of boron the unhybridised p orbital is empty, the development of s character therefore has the effect of deshielding the nucleus, increasing the Lewis acidity at boron. With the Lewis acidity of the boron increased the Lewis acid–Lewis base interaction between boron, and the oxygen within the approaching water molecule is augmented, weakening the adjoining O–H bond and increasing the lability of the acidic hydrogen.

Scheme 10 *To conform to the constraints enforced upon it on complexation, boron adopts geometric characteristics between sp^2 and sp^3 hybridised orbitals, causing the vacant p orbital to develop some s character. This has the effect of deshielding the nucleus and in turn increasing the Lewis acidity at boron. The net effect is to increase the acidity of the labile hydrogen thus making the system more acidic.*

3.2.3 Computational Analysis

Over the period 2003–2004 the relationship between boronic acid diol complexation and acidity was studied computationally. Bock and co-workers modelled the complexation of methanol,[88] 1,2-ethanediol[89] and D-glucose[90] with boronic acids. This approach allowed a quantitative evaluation of the relationship to be undertaken.

The results calculated have been found to broadly match the published experimental data on boronic acid complexation. The value of the O–B–O bond angle calculated for the five-membered ring formed on complexation of a 1,2-diol with dihydroxyborane was 113.8°.[89] A value in good agreement with the O–B–O bond angles observed in the single-crystal X-ray structure subsequently obtained, where values of 113.3° and 113.7° were reported (Figure 7).[86]

The electrophilic reactivity index for the boron atom (ERI_B) of 4-[4-(dimethylamino)[phenylazo]benzeneboronic acid was used to provide a measure of acidity.[90] Examining this ERI_B in a number of bound complexes, the computational results indicated that acidity generally increases on complex formation. Analysis of the two independent boronic acid complexes formed at the 1,2- and 4,6-positions of separate α-D-glucopyranose molecules, revealed a comparative increase in the ERI_B values (relative to the unbound boronic acid) of around 64% for the five-membered ester formed through binding at the 1,2-positions and of around 19% for the six-membered ester formed through binding at the 4,6-positions.

The computational results appear to suggest that considering a proportional relationship between O–B–O bond angle contraction and the acidity of boronic acid when complexed to substrates such as saccharides is somewhat of a simplification. When quantified the measure of boron's acidity was found to be influenced by a number of environmental features such as through space interactions with proximal saccharide oxygen atoms. Nevertheless, in the vast majority of cases the theoretical calculations overwhelmingly agreed with the experimental observations; on complexation of a boronic acid with a polyhydroxyl ligand the O–B–O bond angle decreases and the acidity of the boronic acid increases.

3.3 Complex Formation and Dependence on pH

3.3.1 Empirical Data

From empirical data it has long been clear that pH dominates the reaction kinetics of boronic acid–diol complexation.[80] As discussed above, a step change in the binding constants, of some five orders of magnitude, is apparent at a pH threshold where the pH of the system becomes greater than the boronic acid's pK_a.[72] The kinetics of boronic acid–diol complexation mirror this with an increase in rate of three to four orders of magnitude, when boron is present in its tetrahedral anionic form, rather than as the neutral trigonal species.[91]

When explaining these observations from a mechanistic, arrow pushing perspective the results may appear somewhat counter intuitive. Mechanistically, one could quite reasonably envisage the first step as the attack of a diol oxygen lone pair on boron's vacant *p* orbital,[92] Scheme 11. Yet, experimental data clearly indicates that this reaction is much faster when the *p* orbital is occupied and the boron centre is carrying a negative charge.

Scheme 11 *The expected attack of a diol oxygen lone pair on boron's vacant p orbital does not provide a mechanistic rationale for the four fold acceleration in rate when the boron centre develops a negative charge and the p orbital becomes occupied.*

The value of rate and stability constants are not only dependent on pH; there is also a clear dependence on the acidity of the reacting ligand.[78] While the binding of diols by trigonal boronic acids can be considered to be almost negligible, this is not the case when boronic acids complex with more acidic poly-hydroxyl species such as 1,2-diphenols,[78] α-hydroxy carboxylic acids[70,76–78] and dicarboxylic acids.[75–77,93]

Unfortunately, direct analysis of the reaction parameters of tetrahedral boronates with these acidic ligands is complicated by the presence of reaction pathways, which are kinetically indistinguishable due to proton ambiguity.[79] Therefore, the bias of the literature towards examining the complexation of ligands with neutral boronic acids necessitates that we discuss this first in order to develop a rounded perspective of the mechanism.

3.3.2   Proton Transfer

In work that has been ongoing since the late 1960s, Pizer and co-workers extensively examined the physical properties of trigonal boron acid[§] – ligand complexation.[77,91] From initial studies it was observed that the overall reaction proceeded with a change in geometry at boron, from trigonal to tetrahedral, on complexation.

It was also noted that the rates of these reactions were dependent on a number of features, in particular on the protonation of the ligand (Scheme 12). For the complexation of phenylboronic acid with oxalic acid the rate constant was $k_{COOHCOOH}=2000$ M^{-1} s^{-1}, for the monobasic form the rate dropped to $k_{COOHCOO^-}=330$ M^{-1} s^{-1}, with a rate for the dibasic form of $k_{COO^-COO^-} \leq 0.1$ M^{-1} s^{-1}.[75] These results were unusual for a nucleophilic substitution process in that the highly nucleophilic dianionic species was almost entirely unreactive, while the most protonated attacking species reacted the most rapidly.

The observations were rationalised on the basis of the leaving group expelled and the minimisation of charge repulsion.[75] In the case where a fully protonated ligand was complexed, the more acidic proton would be transferred from the ligand to the hydroxyl on boron to allow the elimination of water.[78] One

---

[§]In keeping with contemporary terminology the term "boron acid" is used within this book as an extension of the nomenclature suggested by Branch *et al.*, where a differentiation is purposely not made between boronic acid and boric acid species.[94,95]

Scheme 12 *While the binding of diols by trigonal boronic acids can be considered almost negligible, this is not the case when boronic acids complex with more acidic poly-hydroxyl species. Here the complexation of trigonal phenylboronic acid with oxalic acid proceeds with a rate constant $k_{COOHCOOH}$ of 2000 M^{-1} s^{-1}.[75] With trigonal boron acids complexation is generally discussed in terms of the overall reaction proceeding with a change in geometry from trigonal to tetrahedral at boron on complexation.*

Key: ---- denotes developing/cleaving covalent bonds

Scheme 13 *The transition state proposed by Pizer was considered associative with the intermediate complex being tetrahedral. In this transition state it was suggested that proton transfer from the entering ligand (here, the oxalic acid on the right) to the leaving hydroxyl (the proton acceptor) could be rate-determining (for clarity the mobile proton is indicated in bold red).[78] One caveat for the proposed transition state was that the interconversion between trigonal and tetrahedral geometries should be rapid, as in the addition of OH^- to $B(OH)_3$, which had been previously assumed to be diffusion controlled.[91]*

proton is therefore an absolute pre-requisite for the reaction. Experimental results indicated that the more acidic proton was involved in proton transfer with a precise correlation between rate and its pK_a. It was postulated that the dependence of reaction rates on the presence of the second, less acidic, proton could be argued on the basis of charge minimisation between the incoming anionic ligand oxygen and the leaving hydroxyl group.

Mechanistically, the pathway was interpreted as a substitution reaction at boron initiated by nucleophilic attack of the ligand on the empty *p* orbital of boron.[75] Pizer proposed a transition state for the reaction of trigonal boron acids with fully protonated ligands. The transition state was discussed in terms of being associative and the intermediate complex being tetrahedral. In this transition state it was suggested that proton transfer from the entering ligand to the leaving hydroxyl could be rate-determining, as in Scheme 13.

To better understand the dependence between rate pH and pK_a, the series of complexes formed from the ligands and trigonal boron acid species depicted in Figures 9 and 10 were examined to establish trends in the stability and rate constant data.[78]

3.3.3 Effect of Altering the Ligand

Across the series of ligands examined in Figure 9 (complexing with the same boron acid) the stability constants for complex formation increased with increasing ligand acidity. This effect was brought about by an increase in the forward rate constant and a decrease in the reverse rate constant for complexation. An observation consistent with the operation of a proton transfer mechanism.

In the forward direction the proton is transferred from the ligand to the hydroxide on boron. In the reverse direction it is transferred from the entering water molecule to the leaving ligand. Therefore, increasing the acidity of the labile proton facilitates the forward proton transfer mechanism, increasing the forward rate of reaction and impedes the reverse proton transfer mechanism, decreasing the reverse rate of reaction.

3.3.4 Effect of Altering the Boron Acid

Across the series of boron acids examined in Figure 10 (complexing with the same ligand) the stability constants for complex formation increased with increasing boron acid acidity. This trend is not a function of proton transfer; instead it reflects a general enhancement in the stability of the product and transition state, a stability that increases with increasingly electronegative

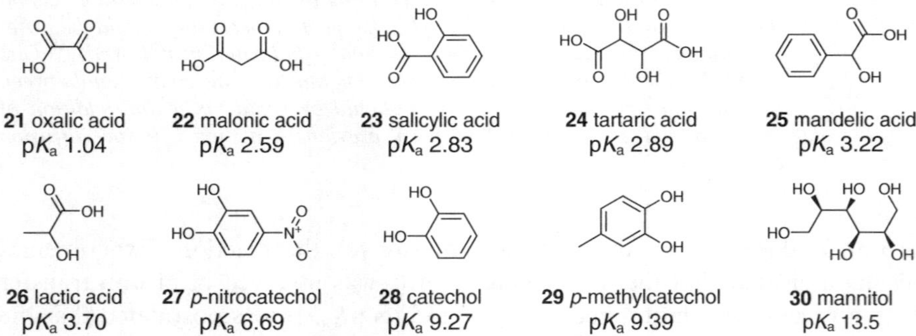

Figure 9 *The series of ligands, with their respective pK$_a$s, as documented by Pizer.[78] It was found that for a given boron acid the stability constants for complex formation increased with increasing ligand acidity.*

Figure 10 *The series of boron acids, with their respective pK$_a$s, as documented by Pizer.[78] It was found that for a given ligand the stability constants for complex formation increased with increasing boron acid acidity.*

substituents. Where acidity was derived from the presence of an electron-withdrawing substituent, the substituent had the effect of delocalising the developing negative charge on boron in the reaction. This stabilised both the transition state and the final anionic product, increasing the stability constant.

Although this accounted for the observed changes qualitatively, in instances where the acidity of the trigonal boron acid was derived in part from the stabilising effect of mixing adjacent π orbitals with the vacant p orbital on boron the trend was non-linear. It was the case that conjugative stabilisation was lost in systems such as phenylboronic acid as soon as the complex became tetrahedral: this change in geometry therefore destabilised the transition state and final ionic product relative to the starting material.

3.3.5 Deuterium Isotope Effect

More recently, a number of other groups have become involved in examining the kinetics of these systems.[96–99] Studies by Funahashi, Ishihara and co-workers examined various factors, such as deuterium isotope effects on the rate of reaction and the reaction activation parameters.[96] There is general concurrence that the first step in the complexation mechanism with trigonal boron acids, the formation of the monoester, is associative and is rate-limiting.[91,97–99] The fact that tetrahedral $RB(OH)_3^-$ reacts much faster than trigonal $RB(OH)_2$ supports the idea that the ring closure should not be rate-limiting if a tetrahedral intermediate is formed. Both research groups agree that proton transfer is significant. However, there are differences in the interpretation of the transition state and rate-determining step.

Funahashi and Ishihara used deuterated reactants to explore complexation and observed a reduction in rate constants of 20–30%.[96] As $k_H/k_D < 2$ this result is in keeping with a secondary kinetic isotope effect, that is to say that while important, it is possible that proton transfer is not the slowest component of the reaction pathway.[99]

From monitoring the reaction in non-aqueous solvents, Funahashi and Ishihara proposed that the interconversion from trigonal to tetrahedral geometry at boron was slow, the formation of the initial B–O bond, not proton transfer, therefore being rate-determining. The proposed plot of reaction coordinate *versus* free energy is illustrated in Scheme 14.[98]

3.3.6 Computational Analysis

To further the mechanistic understanding, computational studies have been conducted by Bock and co-workers, who analysed a range of possible transition states in the formation of the B–O bond. In these studies the reaction was studied on the basis that a molecule of water was eliminated in each step, trigonal geometry was therefore restored to boron in each step. The binding of methanol,[88] 1,2-ethanediol[89] and D-glucose[90] at boron were examined. Where the reactions were simulated *in vacuo* or in acetonitrile the activation barriers were significant. On the other hand, if water, ammonia or NaOH were used in

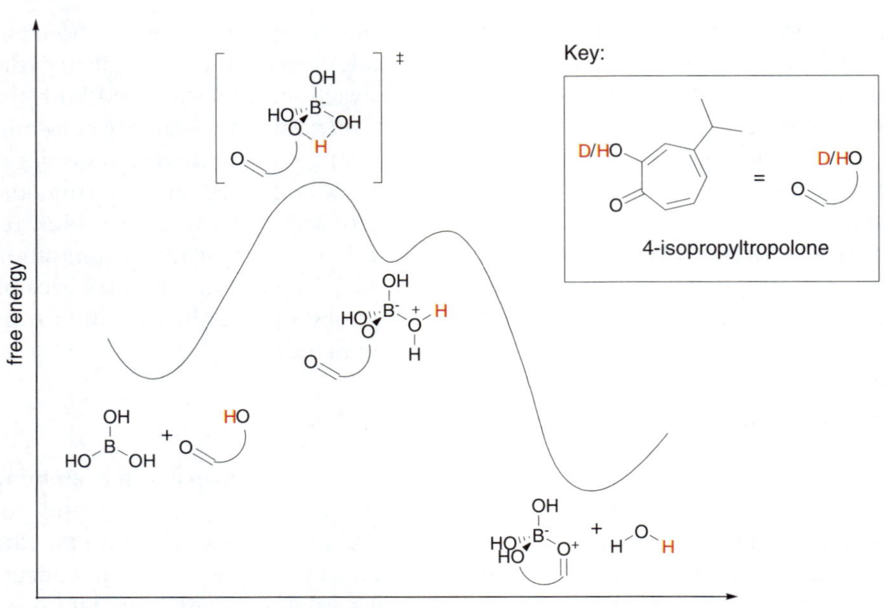

Scheme 14   *Reaction pathway for the complexation of boric acid with 4-isopropyltropolone, based on Funahashi and Ishihara's proposal.[98] This pathway is derived from the results of the deuterium isotope effects as well as the effects of conducting the reaction in non-aqueous solvents and represents the most plausible reaction pathway. In the first step the hydroxyl group of the entering ligand binds to boric acid inducing a change in geometry and forming a tetrahedral intermediate. The hydroxyl proton of the ligand may interact with a hydroxyl group on boron via hydrogen bonding in the transition state to generate the second intermediate with a coordinated water molecule attached. The second step, intramolecular ring closure with dissociation of water, proceeds rapidly.[98] For clarity the hydrogen/deuterium atom under consideration is highlighted in red.*

the capacity of weak Lewis bases the activation barriers were substantially lowered, catalysing the reaction.

The studies ascertained that the initial dehydration step provided the greatest activation barrier, with the proton transfer aspect of the dehydration process being a dominant feature in the transition state.

In vacuo or acetonitrile boron acids reacted *via* strained four-membered transition states, with high activation barriers. In the presence of water, ammonia or NaOH, however, the most stable transition states developed were comprised of six-membered rings, this was achieved by directly utilising a solvent molecule. If water is considered, incorporating an oxygen and a hydrogen atom into the ring, one of boron's hydroxyl oxygen atoms accepts a proton shuttled from a solvent water, while water in turn accepts a proton shuttled from one of the ligand oxygens (Scheme 15).

For the transition state to occur with favourable geometry within the cyclic six-membered transition state a tetrahedral geometry was developed at boron, an O–B–O angle of 107.3° being reported in the presence of water. While the

Scheme 15 *Bock's computational simulation of the lowest energy transition state for 1,2-ethanediol complexing to dihydroxy borane in the presence of water.[90] In water, ammonia and NaOH the most stable transition states developed were comprised of six-membered rings. This was achieved by directly utilising a solvent molecule, allowing a proton to be shuttled via the solvent molecule. Within the cyclic six-membered transition state a tetrahedral geometry was developed at boron, which was required for the transition state to occur with favourable geometry. In the computational system, the rate of proton transfer dramatically increased with the increasing Lewis basicity of the solvent molecules. Also of note is the stabilising hydrogen bond between the lower ligand oxygen and the lower boron hydroxyl depicted in the scheme.*

intermediate is tetrahedral, water is eliminated in this step and the system reverts to trigonal geometry with the formation of a covalent bond between the diol oxygen and boron.

In the computational system, the rate of proton transfer dramatically increased with the increasing Lewis basicity of the solvent molecules. Also of note is a secondary interaction, a stabilising hydrogen bond between the ligand oxygen and the boron hydroxyl that are not actively participating in the bond developing/cleaving aspect of the reaction.

Therefore, while the detailed mechanism by which trigonal boron acid–ligand complexation proceeds has not been fully elucidated, there is general agreement in the literature on a number of salient points.

The reaction proceeds in two distinct steps. The first is bimolecular and is rate-limiting. Nucleophilic attack of a ligand oxygen on the electron deficient trigonal boron forms a complex with tetrahedral geometry.[99] Within this complex a covalent boron–oxygen bond will form, but for this step to proceed a labile proton must be present on the incoming ligand to allow proton transfer to occur.[96] This proton transfer is kinetically significant, occurring either directly or through a solvent chain. The second step is a unimolecular elimination. It is initiated by intramolecular nucleophilic attack at boron by the untethered ligand oxygen. This step closes the ring to form the cyclic diester and to eliminate water.

3.3.7 Reactions with Tetrahedral Borates

Unfortunately, the evaluation of rate constants for the substitution reactions of tetrahedral borates with acidic ligands is complicated by reaction pathways, which are kinetically indistinguishable due to proton ambiguity. There is therefore a comparative paucity of data for complexation with the tetrahedral boronate anions when compared to their neutral trigonal equivalents. As the rates of complexation for tetrahedral borates are much higher than the rates of complexation for trigonal boron acids the only ligands for which the kinetics of tetrahedral borate complexation have been studied are diols (ligands whose protons are not considered acidic in this context), preventing generic trends from being established.

In an effort to circumvent this problem it is possible to consider studies that have been conducted into the binding of boric acid with bidentate ligands in a 1:2 complex.[97,100] By considering the change from a 1:1 (ligand/boric acid) tetrahedral borate complex to a 2:1 (ligand/boric acid) tetrahedral borate complex, the effect of pH on complexation can be studied with no structural change at boron occurring when the pH matches the pK_a.

Where chromotropic acid was studied an increase in acidity of the solution resulted in an increase in the forward and reverse rate constants.[97] Both the forward and reverse reaction rates of reaction were found to be first order with respect to the hydrogen ion concentration. Given the first order dependence on hydrogen ion concentration Yoshimura and co-workers postulated a plausible reaction mechanism for the 1:2 complex formation of boric acid and chromotropic acid, Scheme 16.

Yoshimura's proposed transition state was almost analogous to the one proposed by Pizer for trigonal boron species (page 23) but in the first step the

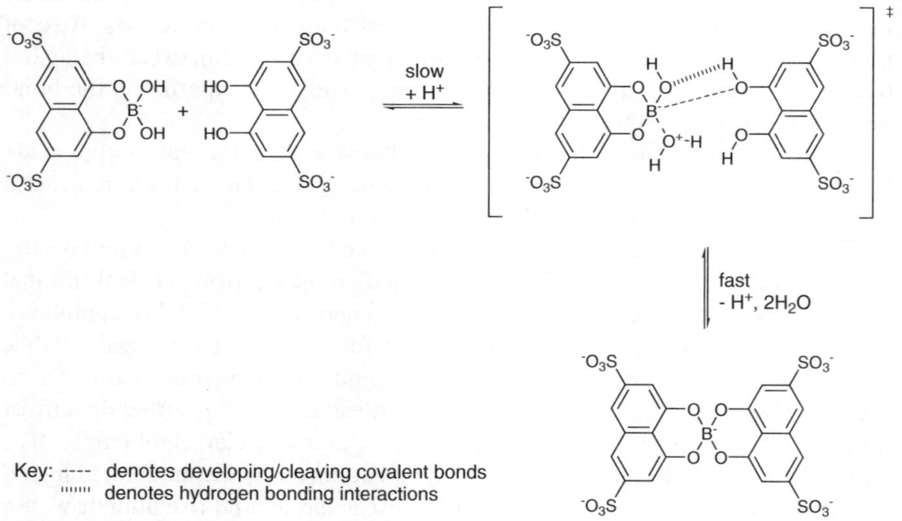

Key: ---- denotes developing/cleaving covalent bonds
'''''' denotes hydrogen bonding interactions

Scheme 16 *As the reaction kinetics of tetrahedral borates with most ligands cannot be followed due to the problems associated with kinetically indistinguishable pathways and proton ambiguity, studies have examined the binding of boric acid with bidentate ligands in a 1:2 complex. By considering the change from a 1:1 to a 1:2 complex a tetrahedral structure is enforced at boron. This allows parallels to be made with the complexation of other tetrahedral boronate anions. Yoshimura's proposed transition state is depicted here, illustrating the complexation of boric acid with chromotropic acid.[97]*

solution protonates a leaving hydroxyl on boron. As the S_N2 reaction proceeds a covalent B–O bond begins to develop, forming a pentavalent boron complex in the transition state. This complex could be stabilised by the formation of a ligand–hydroxyl hydrogen bond (it would also be quite plausible for a proton transfer mechanism to occur *via* a solvent chain with concurrent hydrogen bonding stabilisation, in the manner detailed by Bock). Water would then be eliminated to restore tetrahedral geometry. Nucleophilic attack at boron from the untethered ligand oxygen would re-establish a pentavalent complex, before elimination of water closed the ring and completed the reaction.

3.3.8 Pentavalent Coordination at Boron

A number of transition states with pentavalent coordination at boron have been reported.[101] For example, the reaction of the [BH$_3$–CO] complex with NMe$_3$[102,103] or the hydrolysis of BH_4^-.[104,105] In two instances the hypervalent boron compounds have been found to be stable enough to synthesise and isolate,[106] with Yamashita *et al.* having recently reported the first single-crystal X-ray structure of a pentavalent boron compound, unequivocally confirming the existence of these hypervalent boron species (Figure 11).[107]

3.3.9 Complex Formation and B–O Bond Length Dependence

While unknowns still remain, Pizer has developed a plausible explanation for the observed increase in rate with tetrahedral borates over trigonal boron

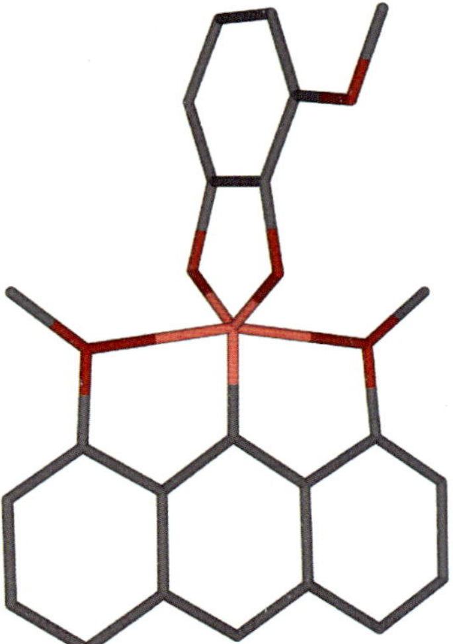

Figure 11 *A transition state with pentavalent coordination at boron would be required for the complexation of a ligand with a tetrahedral boronate anion. Yamashita et al. have recently reported the first single-crystal X-ray structure of a pentavalent boron compound, unequivocally confirming the existence of these hypervalent boron species.[107] Atoms marked in red represent oxygen, pink boron and grey carbon. For clarity hydrogen atoms are not displayed.*

acids.[79] In tetrahedral borates the hydroxyls on boron are likely to be considerably more basic than their trigonal boron acid counterparts. For a good match between trigonal and tetrahedral species it is worthwhile considering the structures of $B(OH)_3$ and $B(OH)_4^-$. Between $B(OH)_3$ and $B(OH)_4^-$ there is a substantial difference in the boron–oxygen bond length, with respective values of 1.37 and 1.48 Å being reported.[85,108]

By augmenting the length and therefore decreasing the strength of the boron–oxygen bond, the hydroxyl on boron becomes more basic, becoming a better proton acceptor and thus facilitating the forward proton transfer mechanism, a mechanism with significant influence on the rates of reaction. While this hypothesis has not been experimentally confirmed or disproved, it does provide a reasonable explanation for the induced step change in reactivity observed when boron acids undergo a change in geometry from sp^2 to sp^3 hybridisation.

3.3.10   B–O Bond Length and Acidification

In reviewing the above systems for this book, it was postulated that the contraction of the O–B–O bond angle on saccharide binding could induce a change in the B–O bond length. It is quite plausible that if a change in the B–O bond length were to occur on binding this may lead to a change in the observed acidity of the system, see Section 3.2.

It would be expected that as the O–B–O bond angle contracted the B–O bond length would increase so as to minimise the effects of any interactions between the two converging hydroxyl groups. An increase in the B–O bond length would, however, induce an increase in the observed basicity of the system; the weaker B–O bond increasing the basicity of the oxygen atom, raising the pK_a (Figure 12).

This notwithstanding, an examination of the single-crystal X-ray structures of the dimeric phenylboronic acid (see Scheme 8)[81] and phenylboronic acid – fructose complex (Figure 7)[86] reveals that there is no change in B–O bond length. In the case of the dimeric phenylboronic acid the O–B–O bond angle is $\sim 116.3°$, this diminishes to $\sim 113.5°$ in the phenylboronic acid – fructose complex, however, for both compounds the average of the B–O bond lengths reported is the same within experimental error, 1.371 Å. Therefore, where trigonal to tetrahedral interconversion is absent, the B–O bond length is not

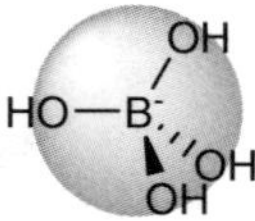

B-O bond length = 1.37 Å   B-O bond length = 1.48 Å

Figure 12   *Rehybridisation of boron from sp^2 to sp^3 is coupled with an increase in the B–O bond length of ~ 0.1 Å. It is postulated that the increased basicity that would accompany such a change would favourably promote the proton transfer process leading to the observed rises in stability and rate constants.*

observed to change on complexation for these two reported structures. This allows us to eliminate a change in the B–O bond length as a likely cause of the change in acidity at boron on boronic acid – ligand complexation.

3.4   Binding Constants and the Influence of Lewis Bases

The final point to be considered in boronic acid – saccharide binding is the predisposition of boronic acids to interact with different kinds of diols. The stability constants (K) between various polyols and boronic acids were first quantified by Lorand and Edwards and it is the case that the trends established now appear inherent in all monoboronic acids.[64]

It is well known that boronic acids readily form five-membered rings with vicinal *cis*-diols in basic aqueous media. It is also the case that six-membered rings can be formed with trans-1,3-diol groups, although the stability of these cyclic diesters is somewhat lower than their five-membered counterparts.[80,109–112]

These trends have been heavily utilised in the design of new sensors and it has been found that the apparent values determined for monoboronic acid sensors concur with the trends of the equilibrium constants reported in Table 1. Apparent values depend heavily on the environments in which the sensors have been studied and are influenced by factors such as: pH, buffer, solvent composition, the concentration of the reactants and the presence of any Lewis basic components.

In response to the growing demand for an accurate interpretation of the complex multi-component equilibria involved in these systems, an extensive investigation was conducted using the potentiometric titration technique.[72] The

Table 1 *The binding constants calculated by Lorand and Edwards for phenylboronic acid in water at 25°C.[64] The four monosaccharides most routinely used in the evaluation of sensors for saccharides are highlighted in* RED.

Polyol	$K\ mol^{-1}$
1,3-Propanediol	0.88
Ethylene glycol	2.80
Propylene glycol	3.80
3-Methoxy-1,2-propanediol	8.50
D-glucose	110
D-mannose	170
D-galactose	280
Pentaerythritol	650
Mannitol	2 300
D-fructose	4 400
Catechol	18 000

refined values for the acidity and binding constants displayed a good match with previously published data. Nevertheless, it became clear from the results that the thermodynamic cycle in Scheme 7 is somewhat of a simplification. Stable complexes were found to form readily between boronic acids and Lewis bases. This represents a significant finding in sensor design as Lewis bases such as the conjugate bases of phosphate, citrate and imidazole are commonly used to buffer the pH of solutions during the spectrophotometric evaluation of sensors, adding a new degree of sophistication to the understanding of the species present in solution.

It was shown that in buffered solutions binary (boronate–Lewis base) complexes as well as ternary (boronate–Lewis base–saccharide) complexes would be formed. Under acidic conditions these ternary complexes are significant and under certain stoichiometric conditions can become the dominant components in solution.

When conducting fluorescence experiments in solutions buffered with a Lewis basic component there is therefore a "medium dependence" related to the formation of Lewis base adducts.[67] These complexes reduce the free boronate and boronic acid concentrations, diminishing the observed stability constants (K_{obs}).

From an experimental perspective this characteristic means that strongly Lewis basic components should be avoided by researchers in the future study of boronic acid appended systems. With care buffers can be chosen so as not to overwhelm the system being investigated. For example, phosphate buffer (pH 7.5) was found not to make a significant contribution to the observed fluorescence intensity or alter significantly the binding constants of the excited state complex under investigation (Scheme 17).

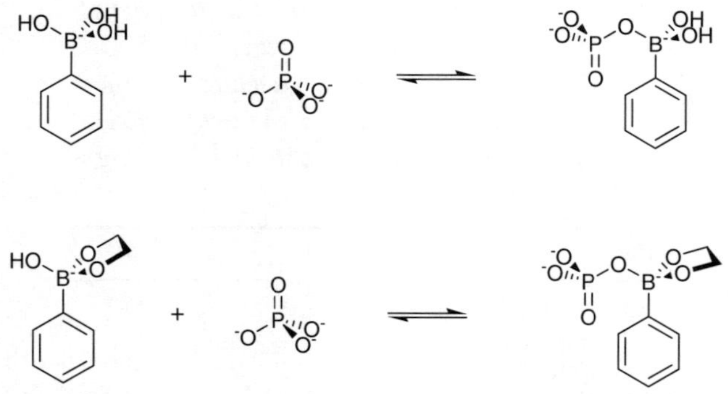

Scheme 17 *In addition to the pair-wise interaction of boronic acids with polyhydroxyl species discussed, boronic acids also form stable complexes with buffer conjugate bases. These complexes can be formed between both the free boronate anion and Lewis bases as well as between saccharide boronate complexes and Lewis bases. Not recognised until 2004, these species persist into acidic solution and under certain stoichiometric conditions can become the dominant component in the solution. The two modes of interaction between the phenylboronate anion and phosphate are illustrated here.*

Therefore, so long as one is aware of the conditions observed stability constants (K_{obs}) have been determined under and the effect that is induced by Lewis basic components in solution, these spectrophotometrically determined constants do provide an accessible, useful and valid measure in the development of boronic acid based sensory systems.

Fluorescent Sensors

The most exciting phrase to hear in science, the one that heralds new discoveries, is not 'Eureka!' but rather, 'Hmm... That's funny...'

Isaac Asimov

4.1   The Application of Fluorescence in Sensing

The use of light to transfer information on events occurring between molecules is particularly advantageous; it provides researchers with a natural interface between events occurring at the molecular level and the macroscopic one. As optical signals convey information through space, fluorescent sensors can be absorbed into dynamic systems such as living tissue and relay information remotely. Sub-millisecond response times are typical, allowing information to be communicated in real time and, if targeted correctly, fluorophores can be located with sub-nanometre accuracy, in effect permitting real-space monitoring.[113] Fluorescence also demonstrates an exceptionally high sensitivity of detection; under controlled conditions, modern instrumentation has allowed analysts to detect responses from single fluorescent molecules[114–116] and in the case of fluorescent sensors, from single guest molecules.[117]

As fluorescent sensors are capable of reporting on a wealth of physical information at low concentrations (micromolar concentrations are typical), they can operate with the minimum of disturbance to the system being investigated. From an analytical perspective, these characteristics are attractive and commercially the parsimonious quantities of compound required can be used to offset synthetic costs.

Fluorescent sensors can be found in many recent analytical advances, such as the continuous monitoring systems developed by immobilising fluorescent sensors onto fibre optic-sensing arrays,[118] or the live imaging of analytes within cells through confocal microscopy.[119] Commercial examples of the use of fluorescent sensors include clinical tools, such as the blood gas analysers that are now commonplace within hospital high-dependency wards and ambulances allowing point-of-care diagnostic monitoring,[120–123] or the glucose-responsive contact lenses currently being pioneered by the Lakowicz research group.[124] These examples highlight the robust and versatile nature of fluorescent sensors, which in turn permit rapid and accurate analyte diagnosis by portable devices.

The use of boronic acids in the development of fluorescent sensors for saccharides is a comparatively young area of research. The first publication on the subject was made by Czarnik in 1992.[125] D-Glucose selectivity was achieved two years later in 1994 by James[126] within Shinkai's industrious research group, who followed this up with enantioselective saccharide recognition a year later.[127] In the intervening decade, the field has blossomed, hundreds of publications on boronic acid–saccharide recognition now embellish the scientific press, with the three articles above having been cited in more than 708 separate publications.¶

Across the diverse range of boronic acid-based fluorescent sensors developed, two distinct design principles predominate in the scientific literature: internal charge transfer (ICT) and PET. In both cases, successful signalling of the binding event arises from alternative low-energy pathways being discretely offered to either the bound or unbound sensors, these processes affecting defined spectral changes in the emission band.

4.2 Photoexcitation and Subsequent Relaxation

Electronic excitation of a fluorophore in its singlet ground state (S_0), by an incident photon of sufficient energy, promotes one of the fluorophore's electrons into a level of higher energy. The excited energy level initially populated will be a singlet state (such as S_1, S_2 or higher), where excitation into a vibrational level within the singlet state is commonplace.[128]

$$\text{Fluorophore} \xrightarrow{h\nu} \text{Fluorophore}^*$$

For most molecules in solution, the subsequent relaxation of the molecule's energy will be rapid. Vibrational energy is swiftly lost through collisional deactivation with solvent molecules and irrespective of the singlet state initially populated, the majority of compounds will follow Kasha's rule;[129] internal conversion rapidly reducing the energy of the molecule non-radiatively until it reaches the lowest excited state available to it, the first singlet excited state (S_1). From this S_1 state (external factors permitting), fluorophores will often dissipate their remaining energy as light. The emission of a photon from this locally excited (LE) state will result in an emission wavelength corresponding to the difference in energy between the initial and final electronic and vibrational energy levels occupied (see Scheme 18).

In considering sensor design, the time dependency of these processes is not insignificant. Typical fluorescence excited-state lifetimes are of the order of 10 ns (1×10^{-8} s).[128] The absorption of light is rapid, electronic excitation occurring in around 10^{-15} s, subsequent internal conversion to the S_1 state is also rapid, occurring in around 10^{-12} s.[128] It is the lifetime of the S_1 state that is comparatively long, accounting for the ~ 10 ns excited-state lifetimes, and it is therefore here that many of the processes responsible for molecular signalling begin.

¶ Source: *ISI Web of Knowledge*, August 2006.

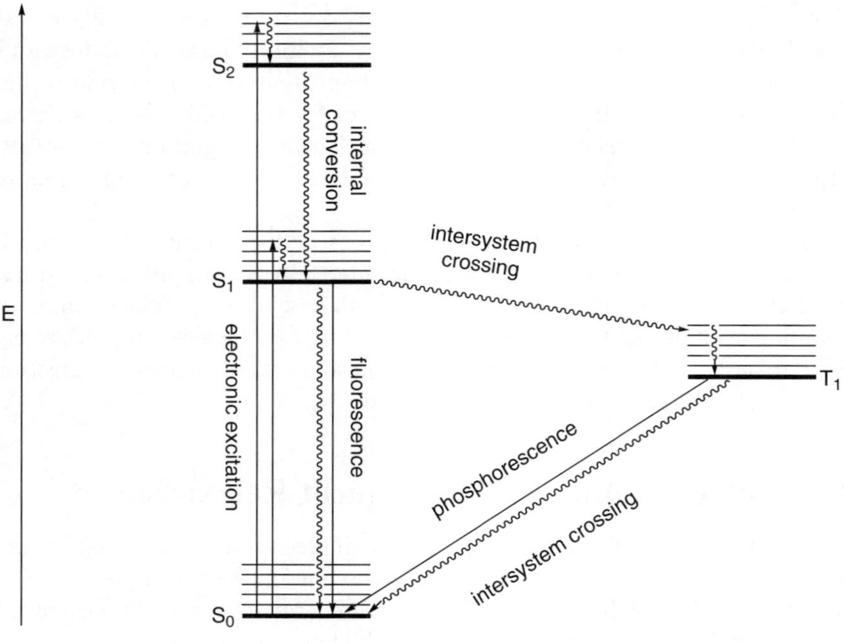

Scheme 18 *A Jabłoński diagram typical of a fluorophore in a LE state with a single emission band. Possible intersystem crossing leading to phosphorescence from the first triplet state (T₁) is illustrated for completeness.*

4.3   Excited State Internal Charge Transfer

4.3.1   Solvent Relaxation

The characteristic Stoke's shift, or tendency of fluorophores to emit light at a longer wavelength than that absorbed, is readily understood by considering the loss of energy due to the internal relaxation processes such as those described in the Jabłoński diagram in Scheme 18. Not only do the surrounding solvent molecules provide a convenient interface through which to dissipate a fluorophore's excess energy to the bulk system but they can also play a key role in lowering the energy of the S_1 state itself.

If a fluorophore with a dipole moment in the ground state is promoted to an electronically excited state, it is generally the case that the strength of the electronic dipole moment will grow.[128] In solutions where polar solvent molecules are free to re-orientate themselves and maximise favourable dipole–dipole interactions, the energy of the excited state can be lowered. For solvents at room temperature, solvent relaxation typically occurs in 10^{-11} s.[128] This is

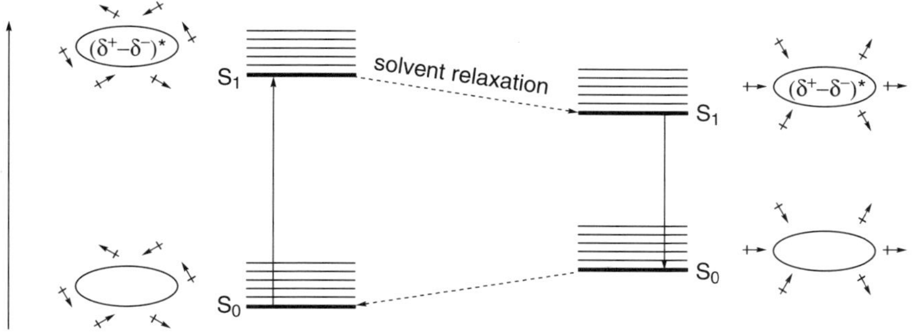

Scheme 19 *The re-organisation of polar solvent molecules around the enlarged dipole a fluorophore in its LE state permits dipole–dipole interactions to lower the energy of the S_1 state, raising that of the S_0 ground state. The stabilising solvent relaxation illustrated will redshift the single emission band of the LE fluorophore.*

too long a time for solvent relaxation to influence the energetics of electronic excitation and the corresponding absorption spectra. Nevertheless, solvent relaxation occurs well within the lifetime of the S_1 state. In instances where this relaxation mechanism is feasible, the energy of the S_1 excited state will be lowered and the energy of the S_0 ground state will be raised, as illustrated in Scheme 19. The net effect of this process is therefore to red shift the emission band of the fluorophore.

4.3.2 Dual Fluorescence

In many respects the stabilisation ICT imparts can be considered as an extension of the solvent relaxation mechanism. In systems where heteroatoms or strongly inductive functional groups are in contact, an imbalance in the electron density can be induced across the excited state of the molecule. This reversible charge separation creates an enlarged dipole moment, enhancing the favourable dipole–dipole interaction of the fluorophore with the surrounding solvent shell. This environmentally dependent stabilisation of the S_1 state is a process, which can lead to dramatic changes in the wavelength of the emission band and as such has found great application in the field of sensing (Scheme 20).

One of the best examples of an ICT system remains the first compound to be clearly identified as altering its fluorescence properties through ICT, *p*-(*N*,*N*-dimethylamino)-benzonitrile **33**.[130] In non-polar solvents, the benzonitrile **33** was observed to display a "normal" fluorescence response, *i.e.*, that characteristic of a benzene derivative in its LE state. In polar solvents, a second emission band of longer wavelength emerged. The relative intensity of the long-wavelength ICT band was seen to grow with decreasing intensity of the short-wavelength LE band as a function of the increasing polarity of the

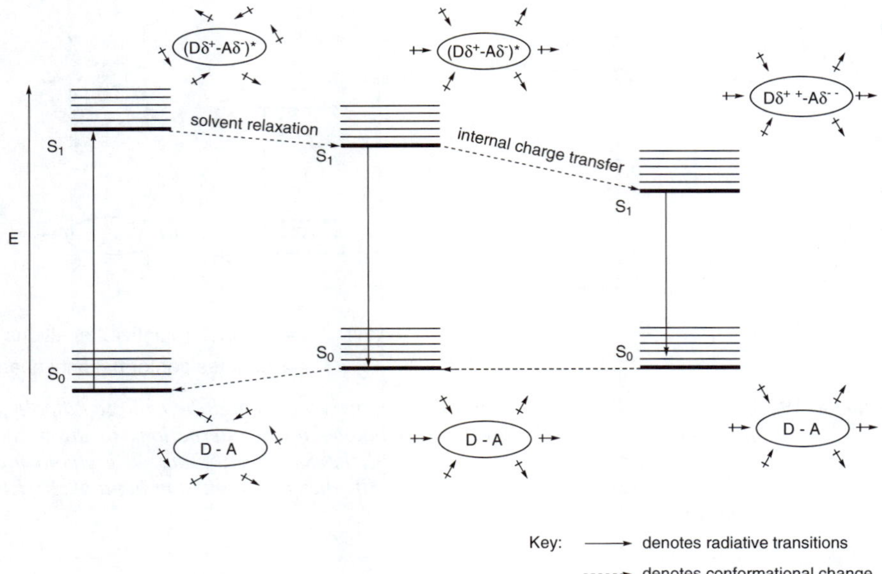

Scheme 20 *The dual emission band of an ICT fluorophore is illustrated. While emission from the LE state may be modestly redshifted, the ICT will affect a dramatic shift in wavelength, environmental conditions allowing. D–A denotes a molecule containing electron donor and electron-acceptor moieties.*

medium; the dual fluorescence being strongly dependent on solvent polarity and temperature.[131]

33

The structure of the benzonitrile **33** is quite typical of ICT systems, which are now described in terms of incorporating an electron "donor" group and an electron "acceptor" group in close proximity or linked through a conjugated π-system.[132]

4.3.3 Excited State Twisted Internal Charge Transfer

Benzonitrile **33** provides a particularly useful example as it illustrates a further process which can occur in ICT systems. Lippert *et al.* determined the value of the dipole moment in the excited ICT state of benzonitrile **33**.[131] In keeping with the discussion above, it was found to be significantly larger than that for the LE state, this finding initially led researchers to believe that the ICT state must resemble a structure approaching the highly dipolar quinoid structure, Scheme 21 (**B**).[133]

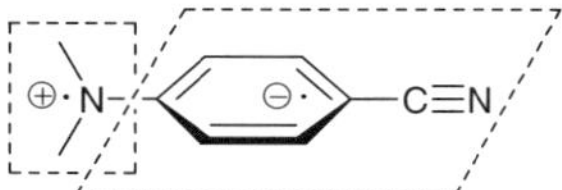

A **B**

Scheme 21 *(A) Expected ground-state structure of benzonitrile **33**. (B) The highly dipolar quinoid structure initially expected for the excited ICT state of benzonitrile **33**.*

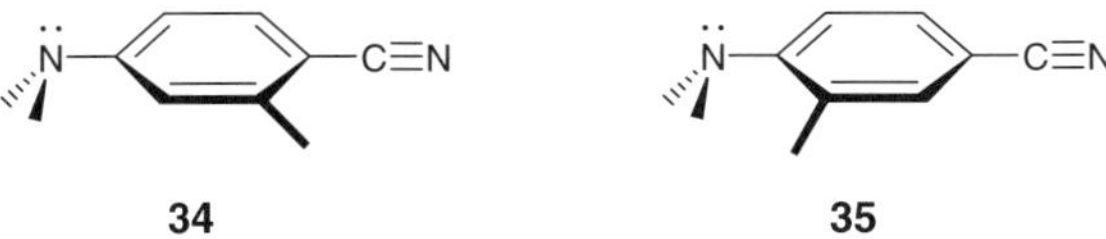

Figure 13 *The dimethylamino group is highly twisted in the CT excited-state conformation, possibly perpendicular to the aromatic ring.[133] This concept is illustrated here with the radical ions of compound **33** in the TICT excited state mutually orthogonally aligned and linked by a single bond.*

A re-investigation of the anomalous properties of benzonitrile **33** by Grabowski and co-workers revealed that this was not sufficient to account for all the experimental data obtained. The properties of the structural derivatives of benzonitrile **33** with methyl ring substituents *meta-* and *ortho-* to the amine group were examined, compounds **34** and **35**, respectively.[134,135] It was found that in the same solvent system, compounds **34** and **35** behaved very differently. The *meta*-derivative **34** displayed dual fluorescence with a dominant band from the LE state, while the *ortho*-derivative **35** displayed fluorescence almost exclusively from the ICT state.

34 **35**

This finding was tentatively attributed to a steric effect, with the LE state being assigned to a coplanar (quinoid-like) structure and the ICT state being assigned to a structure in which the dimethylamino group was perpendicular to the benzene ring. In such a conformation, the orbitals of the amino lone pair and the delocalised π-system of the benzene ring will be orthogonally aligned and no overlap will occur between them. The system can therefore be considered orbitally decoupled and while this situation holds a full electron transfer (ET) can occur (Figure 13).

This phenomenon was found to occur in a number of analogous species, the twisted conformation generally representing the most favourable species. This explanation held for over a decade and as such led to the name of excited state twisted internal charge transfer (TICT).[136] In the early 1970s, new experimental evidence revealed that the TICT model failed to provide a consistent interpretation for empirical data in certain cases, nevertheless the concept of a twisted state still provides a valuable model. For a discussion on the range of hypotheses subsequently postulated regarding the structure of the ICT state and the mechanism and timescale of its formation (the exact mechanistic details still remain a point of contention after 30 years focused research), the interested reader is directed to the extensive review recently published by Grabowski, Rotkiewicz and Rettig.[133]

4.4   Fluorescent Internal Charge Transfer Sensory Systems

In the case of boronic acid-appended fluorescent ICT sensors, the design must consist of a receptor and a fluorophore. Crucially, they must be integrated such that there is significant overlap between the two fragments, as illustrated in Figure 14.

This design generally permits guest molecules to influence the electronic properties of donor or acceptor groups within ICT molecules, or otherwise invoke a conformational restraint in the rotation of key TICT bonds. Although changes in the emission intensity are often observed in parallel with changes in the emission wavelength, the principal designed mode of action for an ICT sensor is to signal substrate binding *via* a shift in the emission wavelength.

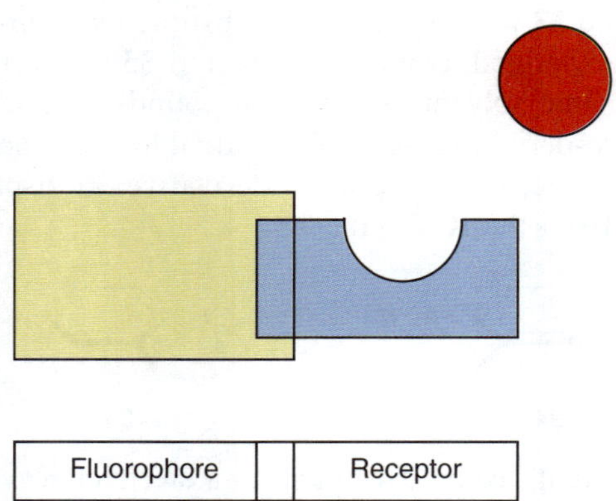

Figure 14   *Schematic representation of the overlapped fluorophore–receptor design assembly for fluorescent ICT sensory systems.*

4.1.1   Early Fluorescent Sensors for Saccharides

The first fluorescent sensor for saccharides was reported by Czarnik in 1992.[125] The sensor (**36**) was comprised of a boronic acid fragment directly attached to anthracene. On addition of saccharide, it was noted that the intensity of the fluorescence emission for the 2-anthrylboronic acid **36** was reduced by ~30%. This change in fluorescence emission intensity is ascribed to the change in electronics that accompanies re-hybridisation at boron. For boronic acid **36** (below its pK_a), the neutral sp^2 hybridised boronic acid displayed a strong fluorescence emission, (above its pK_a) the anionic sp^3 boronate displayed a reduction in the intensity of fluorescence emission (Figure 15).

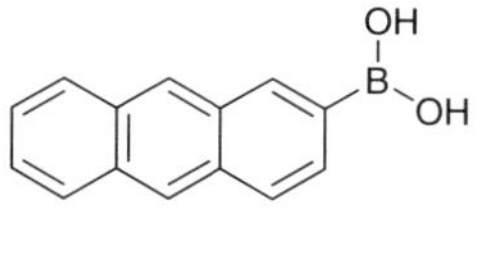

36 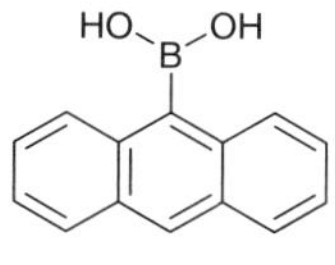   **37**

Figure 15 *The first fluorescent sensor for saccharides: 2-anthrylboronic acid **36**, as well as its isomer 9-anthrylboronic acid **37**. Recognition of saccharides was mediated by an apparent decrease in fluorescence intensity, due to the formation of the boronate anion.*

As discussed in Section 3.2 the formation of a boronic acid–saccharide complex acidifies the boron atom, making the resultant boronic ester more acidic than the initial uncomplexed boronic acid. In this instance, a pK_a of 8.8 was reported for the neutral 2-anthrylboronic acid, and a pK_a' of 5.9 was reported for the 2-anthrylboronic acid complex formed in saturated fructose solution. Exploiting this phenomenon, the system was buffered to a pH of 7.4, a value between the corresponding pK_a and pK_a' values reported. With this constraint in place, a high-fluorescence emission intensity was observed from the uncomplexed boronic acid (pH < pK_a). However, under these buffered conditions, addition of a saccharide to the solution formed the boronic ester, lowering the acidity of the boronic species below the pH of the solution (pK_a' < pH). As a direct result, the boronate anion was generated inducing the decrease in fluorescence observed on addition of saccharide.

The corresponding isomer 9-anthrylboronic **37**, was also examined but displayed smaller changes in fluorescence emission, a feature attributed to the unfavourable *peri*-interactions that would be expected at the 9-position.[137] The observed stability constant (K_{obs}) for boronic acid **36** was 270 M^{-1} with D-fructose in 1:99 (v/v) DMSO/water at pH 7.4 (phosphate buffer).[‖]

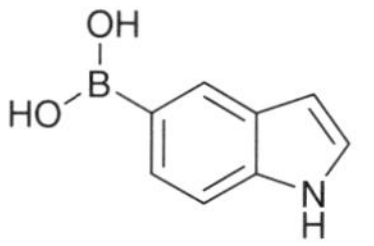

38

The research group of Aoyama investigated the fluorescence properties of 5-indolylboronic acid **38** with mono-, di- and higher saccharides.[139] It was found that the stability constants of the monosaccharides mirrored the

---

[‖] In instances where the stability constants documented in the source publications are dependent on the spectroscopic properties of the evaluated species in solution they are presented here as observed stability constants (K_{obs}) to denote apparent values. Where the stability of the complexes formed was reported in the form of dissociation constants or logarithmically, the values have been converted to (observed) stability constants (K or K_{obs}) within this book to provide consistent and comparable results.[138] To avoid the introduction of mathematical errors through rounding, the values are reported to two significant figures (where appropriate) to match the general level of accuracy found for these values within the scientific literature.

qualitative trend established by Lorand and Edwards for phenylboronic acid.[64] When disaccharides were introduced, however, a great deal of further information was gleaned, the longer-chained saccharides being comparatively more stable when bound to 5-indolylboronic acid **38** than to phenylboronic acid **16**.

It was postulated that the increased stabilisation imparted to complexes with the longer-chained saccharides could be derived from secondary effects such as advantageous hydrophobic interactions with the extended aromatic π-system or from CH-π interactions.

In keeping with the documented results for other arylboronic acids,[109,111,112] it was observed that the reducing sugars examined had far higher observed stability constants (K_{obs}) than the non-reducing sugars examined (devoid of a hydroxyl group on the anomeric carbon). The non-reducing sugars displayed observed stability constants (K_{obs}) values with indole **38**, which were either low or zero.

Further, more subtle trends were also documented. Between linkage isomers it was noticed that there was a selectivity for the 1,6-linked disaccharides over those that were 1,4-linked. More discrete still, was the selectivity between different anomers, a preference being observed for glucose appended disaccharides with an α-configuration over those with a β-configuration.

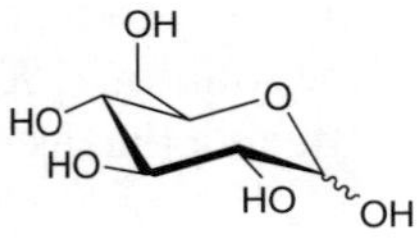

1 D-glucose

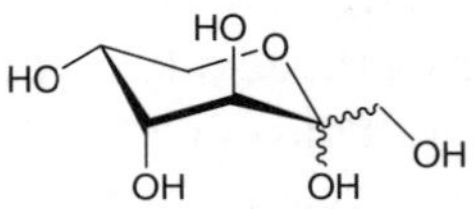

39 D-fructose

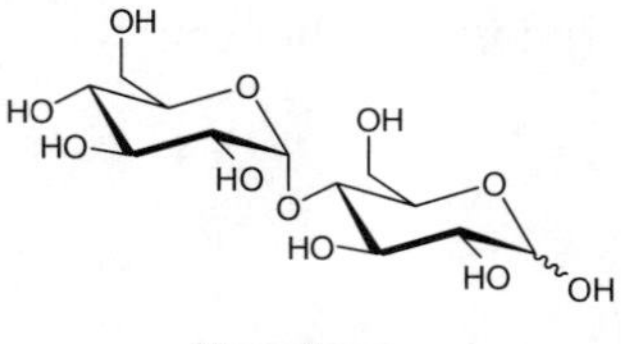

40 maltose
α-D-glucopyranosyl-(1→4)-D-glucose

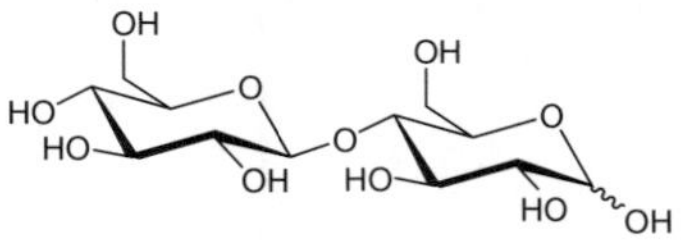

41 cellobiose
β-D-glucopyranosyl-(1→4)-D-glucose

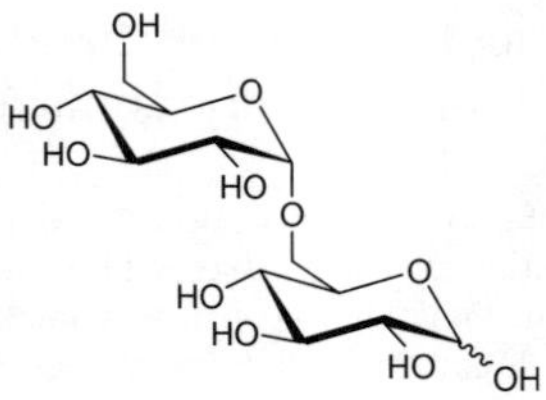

42 isomaltose
α-D-glucopyranosyl-(1→6)-D-glucose

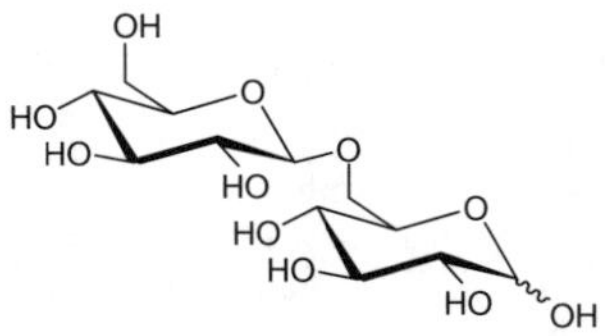

43 gentiobiose
β-D-glucopyranosyl-(1→6)-D-glucose

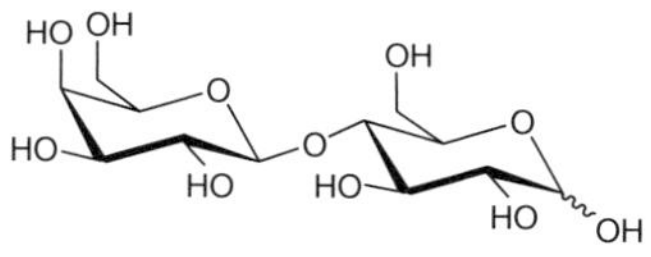

44 lactose
α-D-galactopyranosyl-(1→4)-D-glucose

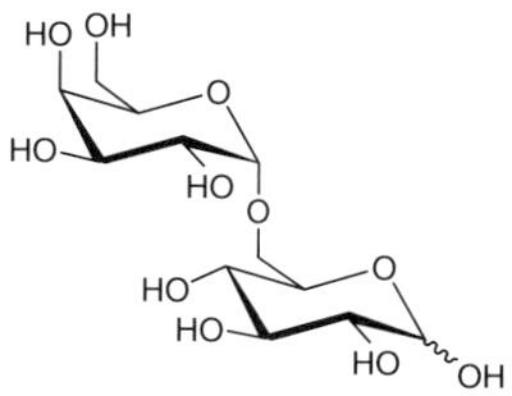

45 melibiose
β-D-galactopyranosyl-(1→6)-D-glucose

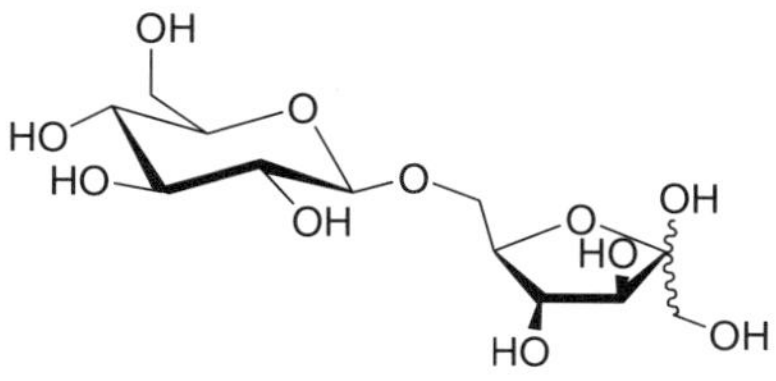

46 palatinose
α-D-galactopyranosyl-(1→6)-D-fructose

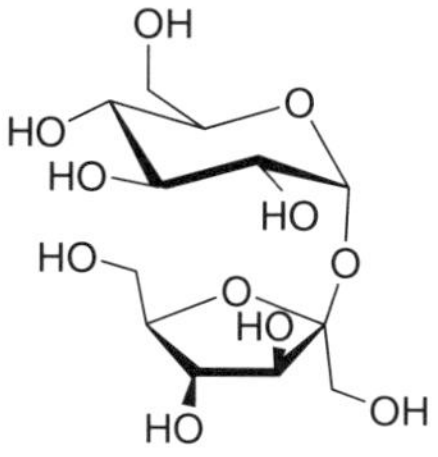

47 sucrose
β-D-fructofuranosyl-(2↔1)-α-D-glucopyranoside

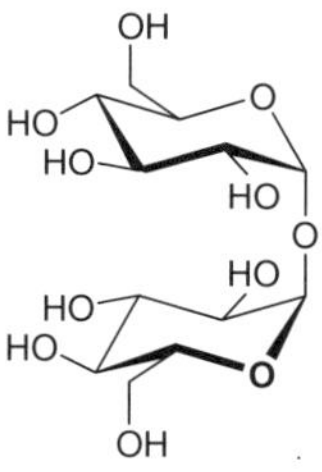

48 trehalose
α-D-glucopyranosyl-(1↔1)-α-D-glucopyranoside

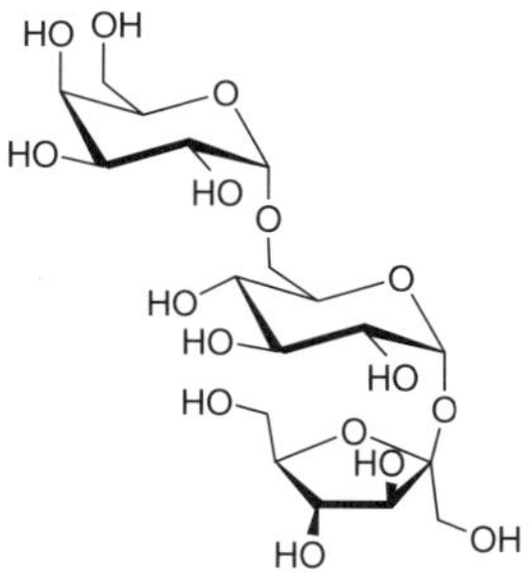

49 raffinose
α-D-galactopyranosyl-(1→6)-α-D-glucopyranosyl-(1↔2)-β-D-fructofuranoside

The observed stability constants (K_{obs}) for indole **38** were 71 M^{-1} with D-glucose **1**, 6 300 M^{-1} with D-fructose **39**, 72 M^{-1} with maltose **40**, 30 M^{-1} with cellobiose **41**, 310 M^{-1} with isomaltose **42**, 250 M^{-1} with gentiobiose **43**, 90 M^{-1} with lactose **44**, 580 M^{-1} with melibiose **45**, 3400 M^{-1} with palatinose **46**, *ca.* 0 M^{-1} with sucrose **47**, *ca.* 0 M^{-1} with trehalose **48**, 290 M^{-1} with maltotriose [α-D-glucopyranosyl-(1→4)-α-D-glucopyranosyl-(1→4)-D-glucose], 670 M^{-1} with maltotetrose [α-D-glucopyranosyl-(1→4)-α-D-glucopyranosyl-(1→4)-α-D-glucopyranosyl-(1→4)-D-glucose], 930 M^{-1} with maltopentose [α-D-glucopyranosyl-(1→4)-α-D-glucopyranosyl-(1→4)-α-D-glucopyranosyl(1→4)-α-D-glucopyranosyl-(1→4)-D-glucose] and *ca.* 0 M^{-1} with raffinose **49** in water.

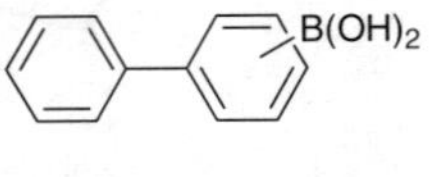

50: 2-B(OH)$_2$,
51: 3-B(OH)$_2$,
52: 4-B(OH)$_2$

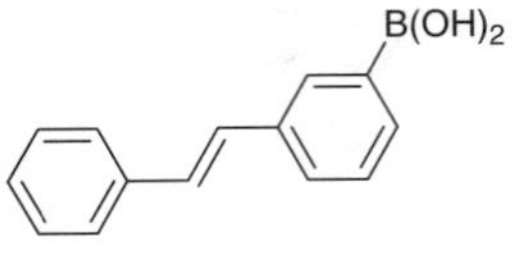

53

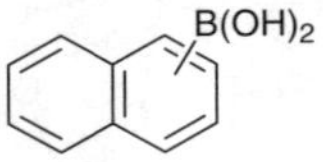

54: 1-B(OH)$_2$,
55: 2-B(OH)$_2$

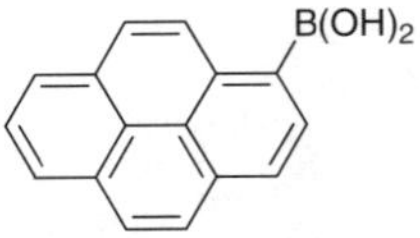

56

These systems were further investigated by Shinkai and co-workers who examined sensors **36**, and **50–56**. The compounds were evaluated against D-fructose and it was found that sensors **51** and **55** functioned particularly effectively.[140,141] The sensors displayed strong fluorescence emission coupled with far greater fluorescence quenching on saccharide binding than previously observed (reductions in fluorescence emission of 70% and 82% were observed for sensors **51** and **55**, respectively). Additionally, a large induced change in the acidity of the boron species was observed on complex formation. These three characteristics made sensors **51** and **55** particularly good candidates for saccharide sensors. Indeed a strong fluorescence emission, large fluorescence response and large-induced shift in the acidity of the boron species are now considered essential for the efficiency of many contemporary boronic acid appended fluorescent sensors.

Expanding on the work of Shinkai and co-workers,[142] DiCesare and Lakowicz examined a range of stilbene boronic acid derivatives to further understand the simple boronic acid–fluorophore systems discussed above.[143,144] It was initially believed that fluorescence quenching from the boronate anion through PET caused the fluorescence response of sensors **36–56**. A more reasonable explanation was provided by examining the electronic properties of the boronic acid stilbene derivatives with electron withdrawing and electron donating groups introduced at the 4′ position.

The interpretation of these results indicates that changes in electronic properties of the boron group, not PET, influence the spectral changes of the fluorophores; accordingly, the systems are now generally discussed as ICT sensors and are treated as such within this book.[9]

57: R = H,
58: R = OMe
59: R = N(Me)$_2$
60: R = CN

61: n = 2
62: n = 3

It was demonstrated that the neutral sp^2 hybridised boronic acid had acceptor group properties and the anionic sp^3 hybridised boronic acid could act as a donor group. In instances, when the moiety at the $4'$ position and the hybridisation at boron conspired to produce a system with a donor and an acceptor linked through the conjugated stilbene scaffold, ICT was found to occur, lowering the energy of the excited state. This influence was elegantly exemplified by the examination of two diametrically opposed systems.

electron donor   electron acceptor

59

In the first case, $4'$-(dimethylamino)stilbene-4-boronic acid **59**, the electron-donating dimethylamino moiety is the donor group. When boron is sp^2 hybridised, and therefore an acceptor, excited-state ICT can occur between the amino donor and boron acceptor, redshifting the emission wavelength of the sp^2 species.

On re-hybridisation of boron to sp^3, its acceptor properties are lost. This leads to a loss of the ICT effect in the excited state of the sp^3 species and shifts the emission wavelength of the fluorophore to higher energy. The inability of the sp^3 hybridised species to lower the energy of its excited state by a mechanism available to the sp^2 hybridised species causes a dramatic change in the properties of the emission band. On sp^2 to sp^3 interconversion a 45 nm blueshift is induced in the emission wavelength coupled with an increase in the emission intensity (Scheme 22).

The fluorescent response to the loss of the electron-withdrawing properties of boron on conversion from sp^2 to sp^3 hybridisation was verified by increasing the pH of a solution of sensor **59** (from 6.0 to 12.0) and by the addition of saccharide to a solution of sensor **59** buffered at pH 8.0 (pK_a'=6.61, pH=8.0, pK_a=9.14), both titrations yielding the same results.

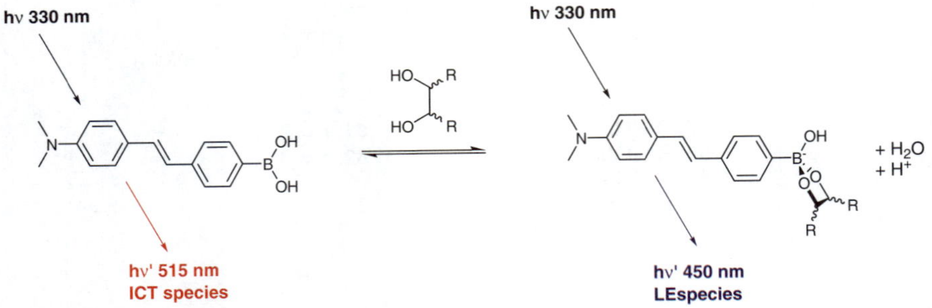

Scheme 22 *On addition of saccharide (or an increase in pH), the overall emission wavelength of sensor* **59** *blueshifts due to disruption of the ICT state. Colours are used to depict relative red and blueshifts in the emission wavelength rather than the actual colour of the emitted light.*

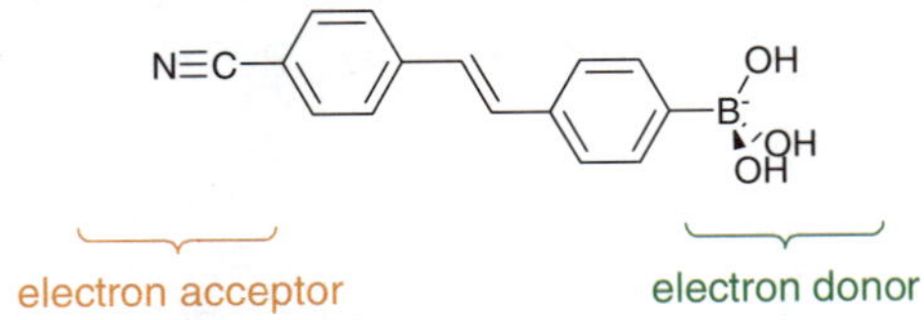

60

Conversely, in 4′-cyanostilbene-4-boronic acid **60**, the electron-withdrawing cyano moiety is the acceptor group. When boron is sp^2 hybridised and therefore also an acceptor, no excited-state ICT is feasible. On re-hybridisation of boron to its sp^3 form, the boron becomes a donor group, allowing ICT to occur between the boron donor and cyano acceptor, redshifting the emission wavelength of the sp^3 species.

Reversing the roles of the donor and acceptor groups for the boronic acid and 4′ moiety yielded values for the changes in emission wavelength and intensity for sensor **60** that were similar in magnitude to sensor **59** but occurred towards opposite ends of the electromagnetic spectrum. On sp^2 to sp^3 inter-conversion a 40 nm redshift was induced in the emission wavelength coupled with a decrease in the emission intensity (Scheme 23).

The fluorescence response to the revival of the electron-donating properties of boron on conversion from sp^2 to sp^3 hybridisation was again verified by increasing the pH of a solution of sensor **60** (from 6.0 to 12.0) and by the addition of saccharide to a solution of sensor **60** buffered at pH 8.0 (pK_a'=5.84, pH=8.0, pK_a=8.17), both titrations yielding the same results.

The observed stability constants (K_{obs}) for sensor **57** were 290 M^{-1} with D-fructose and 91 M^{-1} with D-glucose, for sensor **58** they were 1000 M^{-1} with D-fructose and 23 M^{-1} with D-glucose, for sensor **59** they were 400 M^{-1} with D-fructose and 10 M^{-1} with D-glucose and for sensor **60** they were 1500 M^{-1} with D-fructose and 55 M^{-1} with D-glucose in 2:1 (v/v) methanol/water at pH 8.0 (phosphate buffer).[143,144]

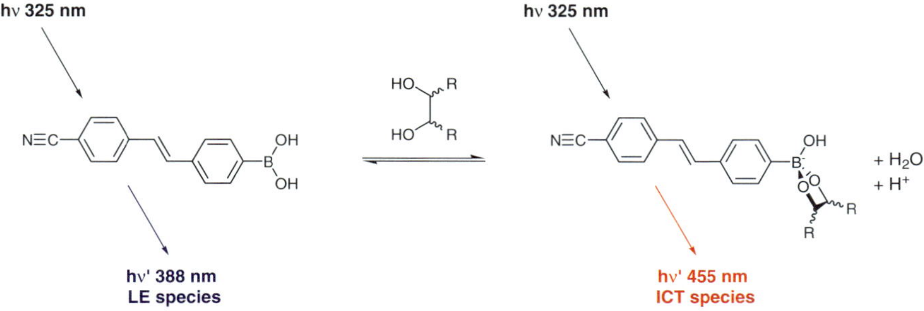

Scheme 23 *On addition of saccharide (or an increase in pH), the overall emission wavelength of sensor* **60** *blueshifts due to disruption of the ICT state. Colours are used to depict relative red- and blueshifts in the emission wavelength rather than the actual colour of the emitted light.*

63   **64**   **66**
               65

Lakowicz quickly recognised the importance of stilbene boronic acid **59** and has since prepared a number of analogous ICT fluorophore systems, including oxazoline[145] **63**, chalcones[146] **64** and **65** and boron-dipyrromethene (BODI-PY)[147] **66**. The observed stability constants (K_{obs}) for **63** were 526 M^{-1} for D-fructose and 27 M^{-1} for D-glucose in water/methanol 2:1 (v/v) at pH 7.0 (phosphate buffer). The observed stability constants (K_{obs}) for **64** and **65** were 400, 476 M^{-1} for D-fructose and 29, 33 M^{-1} for D-glucose in 2:1 (v/v) water/methanol at pH 6.5 (phosphate buffer). The observed stability constants (K_{obs}) for 66 were 1000 M^{-1} for D-fructose and 13.7 M^{-1} for D-glucose in water at pH 7.5 (phosphate buffer).

The oxazoline **63** and chalcone **64** and **65** systems produce large fluorescence changes, however the BODIPY system displays only a small change in fluorescence. Although the BODIPY system **66** did not work particularly well, this class of fluorophore does require further exploration. The BODIPY chromophore possess many advantages as a fluorescent probe, for example, it possess high-extinction coefficients, high-fluorescence quantum yields, good photostability, a narrow emission band and their building block synthesis allows the development of many different analogues showing emission maxima from 500 to 700 nm. Long-wavelength fluorescent probes for D-glucose are highly desirable for transdermal D-glucose monitoring and/or for measurements in whole blood. In addition, narrow emission bands are desirable for high signal to noise ratios. Lakowicz and Geddes have recently demonstrated the usefulness of this type of fluorescent sensor by preparing contact lenses doped with **60**, **59**, **61**, **64** and **65** in order to prepare non-invasive monosaccharide sensors.[148]

67   **68**   **69**   **70**

Wang has recently shown that a very simple naphthalene system **67** can produce very large fluorescence changes (41-fold at 445 nm on addition of 50 mM fructose).[149] The observed stability constants (K_{obs}) for **67** were 207 M^{-1} for D-fructose and 4.0 M^{-1} for D-glucose in water at pH 7.4 (phosphate buffer). The isomeric naphthalene system **68** is a ratiometric sensor that show large fluorescence intensity changes at two wavelengths, 433 nm (increase in intensity) and 133 nm (decrease in intensity).[150] The observed stability constants (K_{obs}) for **68** were 311 M^{-1} for D-fructose and 3.6 M^{-1} for D-glucose in water at pH 7.4 (phosphate buffer). Another isomeric naphthalene **69** has also been investigated and also shows a large decrease in fluorescence (80% at 432 nm on addition of 50 mM fructose).[151] The observed stability constants (K_{obs}) for **69** were 120 M^{-1} for D-fructose and 2.4 M^{-1} for D-glucose in water at pH 7.4 (phosphate buffer). The dibenzofuran-4-boronic acid **70** is interesting in that changes in fluorescence intensity at three wavelengths are observed 301 nm (increase in intensity), 318 nm (increase in intensity) and 327 nm (decrease in intensity).[152] The observed stability constants (K_{obs}) for **70** were 514 M^{-1} for D-fructose and 0.6 M^{-1} for D-glucose in water at pH 7.4 (phosphate buffer).

4.4.2 Fluorescent Internal Charge Transfer Sensors Incorporating the Ortho-(Aminomethyl)Phenylboronic Acid Fragment

Given the importance of receptors incorporating N-methyl-o-(aminomethyl) phenylboronic acid fragments. It seems pertinent to extend the examination of ICT sensors to include those with o-(aminomethyl)phenylboronic acid receptors directly coupled to fluorophores (Figure 16).

The first concerted effort to couple donor and acceptor fragments to produce an ICT sensor for saccharides was by Sandanayake, sensor **71**.[153] In this case coumarin was used as the fluorophore, which led to the redshifting of the emission wavelength and a decrease in the emission intensity on saccharide binding. Unfortunately, these changes were rather modest with an observed stability constant (K_{obs}) of 27 M^{-1} for D-fructose in 1:1 (v/v) methanol/water.

Figure 16 *o-(Aminomethyl)phenylboronic acid derivatives.*

71

72   **73**   **74**

Bosch *et al.* prepared the aniline based sensors **72–74**.[154,155] On saccharide binding, the emission wavelength of sensor **72** was significantly altered. In an aqueous solution, buffered at pH=8.21 the emission maxima of sensor **72** was blueshifted by 42 nm on addition of D-fructose. This observed dual fluorescence was ascribed to TICT in the case of neutral boronic acid **72** and LE states in the case of anionic boronate **72** (Scheme 24).

TICT arises in the neutral boronic acid **72** owing to the uncomplexed structure containing a nitrogen–boron (N–B) interaction. This interaction has been confirmed by X-ray crystallography, the nitrogen lone-pair coordinating to the boron with the N–B bond perpendicular to the anilino π-system. (In protic systems, the direct N–B bond is replaced by the solvent inserted system see Section 4.9) This stabilisation of the excited state is lost as soon as the boronic acid is converted to the anionic boronate, either through the pair-wise complexation of a saccharide guest or through an increase in the pH of the solution.

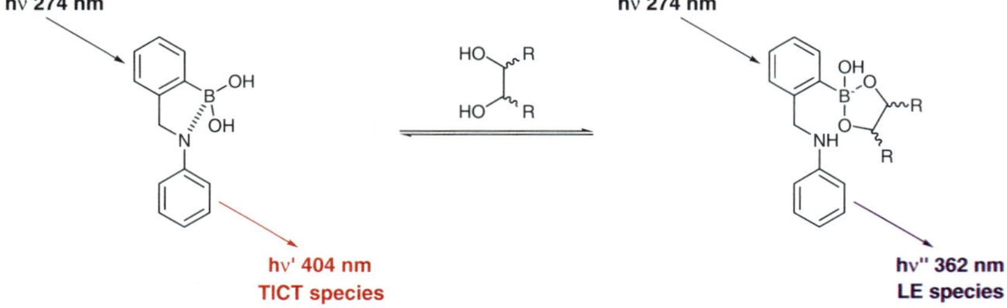

Scheme 24 *Red-shifted fluorescence emission due to TICT of the neutral sp^2 hybridised form of sensor **72** and blue-shifted emission due to the LE sp^3 species; in this case, the boronate anion is formed due to saccharide complexation. Colours are used to depict relative red- and blueshifts in the emission wavelength rather than the colour of the emitted light.*

Evaluation of sensors **73** and **74** revealed that due to the increased N–B distance associated with the *meta-* and *para*-substituted boronic acids no N–B interactions were possible, excluding stabilisation from occurring *via* TICT. The observed stability constants (K_{obs}) for sensors **72**, **73** and **74** were 79, 212 and 129 M^{-1}, respectively, with D-fructose and 6, 8 and 6 M^{-1}, respectively, with D-glucose in 52.1 wt% methanol in aqueous phosphate buffer at pH 8.21.

75

This design was extended to produce sensor **75**, a sensor with D-glucose selectivity induced through the introduction of a second, correctly placed, boronic acid recognition site. As we shall discuss later, the careful orientation of two boronic acid units is pivotal in the design of sensors with selectivity for specific saccharides. In this instance, the observed stability constants (K_{obs}) for sensor **75** were 55 M^{-1} with D-fructose and 140 M^{-1} with D-glucose in 52.1 wt% methanol in aqueous phosphate buffer at pH 8.21.[156]

4.5 Excited State PET

4.5.1 Electron Transfer

ET plays a central role in the physical, chemical and biological sciences. Providing the key steps in photosynthetic and respiratory pathways, as well as many chemical reactions, ET has been extensively studied.[157,158]

Marking the first step in photosynthesis, PET is fast, extremely efficient and can occur over long distances within biological systems; in bacteriochlorophyll the first ET occurs in 3 ps, with 100% quantum yield and over a distance of 17 Å.[159] Extensive biomimetic syntheses have been conducted to further understand these properties and have led to a range of compounds which can actively model this behaviour. To function, the compounds must incorporate proximal electron-donating and electron-accepting groups that are suitably insulated from one another.[160]

In supramolecular compounds, a saturated hydrocarbon bridge has proved effective at insulating the resulting charge-separated species giving rise to a general Donor-Bridge-Acceptor (D-B-A) construction assembly.[157] The length of the bridge is significant, within these systems it is generally the case that the longer the bridge, the less favourable the intramolecular ET process becomes.[161] While long-range intramolecular ET has been elegantly illustrated in numerous elongated, structurally rigid organic molecules such as compound

76,[162] these molecules represent the upper bounds of donor–acceptor separation within synthetic intramolecular ET systems.[163]

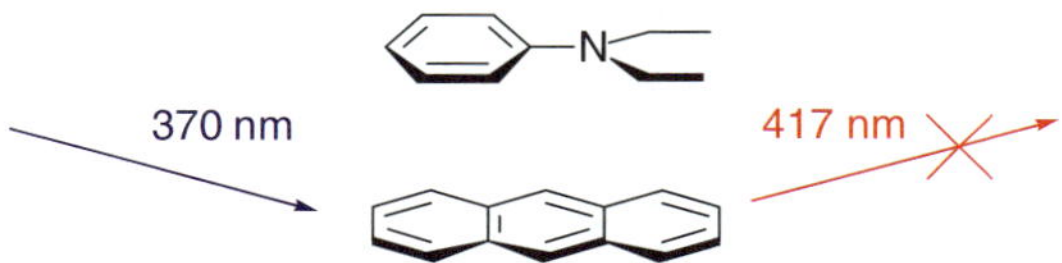

76

In the field of fluorescence, the effect of quenching by ET has long been known.[128] Possibly, the most efficient fluorescence quenchers to operate *via* known ET mechanisms are amines.[128] Amines efficiently inhibit fluorescence emission of most unsubstituted aromatic systems, a good example being the quenching of anthracene's fluorescence by diethylanaline,[164] Scheme 25.

The efficiency of quenching in the Scheme 25 was found to be a function of the solvent polarity (the intermediate radical ion-pair benefiting from increased solvent polarity) and the distance between the donor and acceptor (a maximum encounter distance of ~ 7 Å being reported). From the viewpoint of designing intramolecular ET into supramolecular compounds, the importance of reduced distances between donor and acceptor fragments makes an sp^3 hybridised methylene bridge ideal for ET; the bridge being suitably small, yet capable of effectively insulating the charge-separated species.

This knowledge has been adapted to allow molecular switches to be designed into fluorescent sensors through the use of amine donor groups proximal to suitable fluorophore acceptor groups, separated by methylene bridges. This design principle was first applied to a sensory system by de Silva *et al.* who reported the pH sensor **77**.[165]

Scheme 25 *Amines quench the fluorescence of most unsubstituted aromatic systems. Blue and red are used to depict the relative Stoke's shift in wavelength rather than the observed colour.*

77

The pK_a of the tertiary amino group in sensor **77** is 7.7 in 8:2 water/methanol. Above this pK_a, where the amine is predominantly present as its free base, the amine lone pair can participate in ET and a highly efficient quenching of the fluorescence emission is observed. When the pH is lowered below the pK_a of the amine, so as to protonate it, the lone pair becomes involved in a bonding interaction, disrupting ET and resulting in a complete fluorescence revival.

This molecular construct has been found to function for a large array of systems.[3] The compounds relay information on whether the receptor's binding site is occupied or not by effecting a marked change in the intensity of the fluorescence emission, inducing so called "on" (high fluorescence) or "off" (low fluorescence) emission states of the fluorophore, while retaining many of the intrinsic properties of the receptor and fluorophore moieties.

4.5.2 The Mechanistic Interpretation of PET

Following electronic excitation of the fluorophore and the subsequent relaxation processes that lead to the lowest vibrational level of the S_1 excited state (see Section 4.2), systems in the S_1 D-B-A* state can rapidly lower their energy by undergoing a radiationless** intramolecular ET, see Scheme 26.[3,169,170]

Given the importance of ET this step has been comprehensively studied and a number of sophisticated theories developed to model it.[157,158] Most hinge on the seminal work by Marcus, which was recognised in 1992 with the award of the Nobel prize in chemistry.[2]

As no bond breaking or forming occurs, Marcus theory permits a simple description of the reaction coordinates in terms of the known properties of the reactants and products.[169] The energy of the charge-separated species in the S_1 D$^+$-B-A$^-$ state can therefore be considered as lowered by an amount dependent on the stabilisation derived from the internal rearrangement of bond angles and lengths as well as the rearrangement of the surrounding solvent shell. In compounds where PET can occur the prerequisite is that the energy of the intermediate S_1 D$^+$-B-A$^-$ state is such that it bridges the S_1 D-B-A* and S_0 D-B-A states; the interdigitated vibrational levels allowing internal conversion to occur from the lowest vibrational level of one electronic state to the hot (upper) vibrational levels of the next.[167,169]

$$D\text{-}B\text{-}A \xrightarrow{\ h\nu\ } D\text{-}B\text{-}A^* \xrightarrow{\ ET\ } D^+\text{-}B\text{-}A^-$$

Scheme 26 *Following excitation by an incident photon donor–bridge–acceptor systems can rapidly lower their energy via ET.*

**It is well known that an accelerating charged particle will generate electromagnetic radiation.[166] The emission of radiation at radiofrequency (tetrahertz) wavelengths during intramolecular ET has recently been documented and applied in the measurement of charge transfer rate constants. ET is not therefore a non-radiative process in the strictest sense, however, no emission will be observed in or near visible wavelengths, nor does this radiation play a significant role in the dissipation of the system's energy.[167,168]

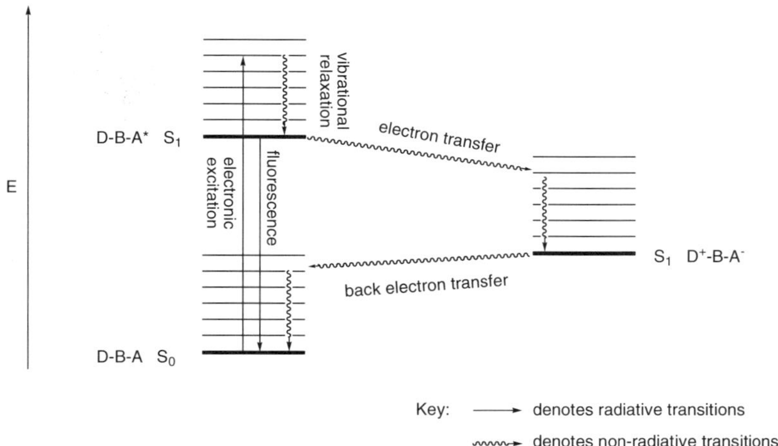

Scheme 27 *Jabłoński diagram depicting the non-radiative pathway available to compounds, which undergo internal PET.*

The back ET†† geminate recombination from the newly formed S_1 D^+-B-A$^-$ charge-separated species to the S_0 D-B-A ground state occurs as an internal conversion process. This rapidly neutralises the potentially damaging imbalance of charge across the compound and restores the molecule to its lowest energy ground state, thus providing the highly efficient mechanism for PET to quench fluorescence, Scheme 27.[167,169,173]

For further information and alternative mechanistic interpretations of PET, the reader is directed to several pathways that have been lucidly described within the literature.[3,167,169,170]

4.6 PET Sensory Systems

In the case of fluorescent PET sensors, the design is often discussed in terms of a receptor and a fluorophore separated by a spacer so as to match the Donor–Bridge–Acceptor motif (Figure 17). To permit the recognition of saccharides *via* boronic acid complexation, the interaction between *o*-methylphenylboronic acids (Lewis acids) and proximal tertiary amines (Lewis bases) has been exploited, Figure 18.

While elucidating the precise nature of the amine base–boronic acid (N–B) interaction has been debated (Section 4.9), it is clear that an interaction does exist and it provides two distinct advantages.

The first, proposed by Wulff,[174] is that the interaction between a boronic acid and proximal amine lowers the pK_a of the boronic acid. This effect is sufficient so as to allow binding to occur at neutral, *i.e.*, physiological, pH.

---

†† In the nomenclature used within this book "back ET" (also commonly called charge recombination[161,171]) denotes the thermal recombination of charge from the S_1 D^+-B-A$^-$ state to the S_0 D-B-A state, so as to restore the donor and acceptor to their original ground state oxidation levels.[172] When examining the scientific press the reader is advised that back ET can been used to denote other ET mechanisms such as the potential S_1 D^+-B-A$^-$ to S_1 D-B-A* pathway.[161]

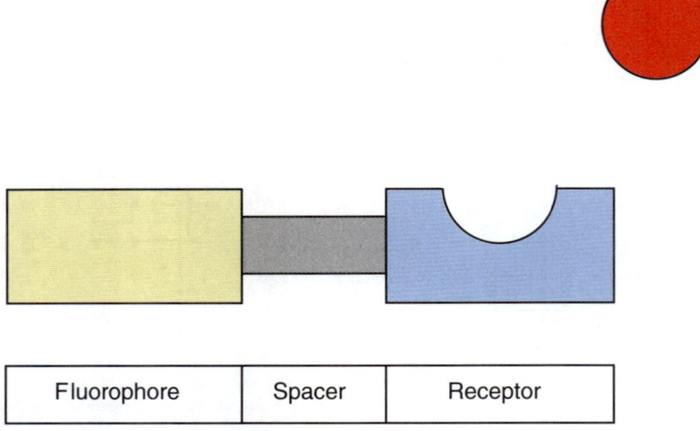

Figure 17 *Schematic representation of the fluorophore–spacer–receptor design assembly for fluorescent PET sensory systems.*

Figure 18 *The generic design of fluorescent PET sensors with the N-methyl-o-(aminomethyl)pheny boronic acid recognition unit.*

The second, relates to the contraction of the O–B–O bond angle of a boronic acid on complexation with a saccharide and the associated increase in acidity at the boron centre (discussed in Section 3.2). This increase in acidity of the already Lewis acidic boron augments the N–B interaction and in turn disrupts PET. The reduction in pK_a at boron on saccharide binding therefore has the net effect of modulating fluorescence emission intensity, *via* the amine group, and introducing a digital "off–on" response from the fluorophore, indicative of the boronic acid receptor being unbound or bound, respectively (see Section 4.9, for further discussion).

The first fluorescent PET sensor for saccharides to employ the *N*-methyl-*o*-(aminomethyl)phenylboronic acid fragment was reported by James *et al.* in 1994.[175,176] Illustrated in Figure 19, sensor **78** functioned remarkably well, a large increase in fluorescence on addition of saccharide was observed, as well as the capacity to function over a broad range of pH. Noticeably, the monoboronic acid **78** displayed the same inherent trend in selectivity for saccharides as had been documented by Lorand and Edwards for phenylboronic acid 35 years earlier, a trend which appears to be inherent to all monoboronic acid receptors of this type. Qualitatively, binding constants (*K*) with monoboronic acids increase in the order D-glucose < D-galactose < D-fructose.[64]

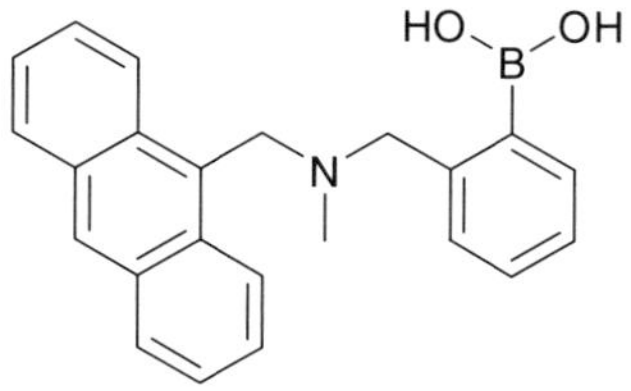

78

Figure 19 *The first fluorescent PET sensor for saccharides to have been rationally designed. In this illustration, the fluorophore–spacer–receptor construction is readily apparent; anthracene represents the fluorophore; methylene, the spacer; and N-methyl-o-(aminomethyl)phenylboronic acid the receptor.*

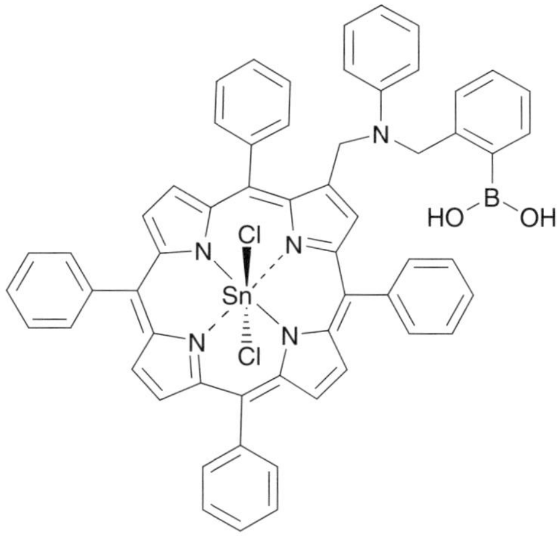

79

Kijima has developed an "on—off" PET system **79**, which behaves in opposition to the conventional "off—on" PET system.[177] In this system, steric crowding on saccharide binding breaks the N–B bond found in the "free" receptor.

4.6.1 Diboronic Acid Sensory Systems with Selectivity for Specific Saccharides

The D-fructose selective monoboronic acid-based sensor **78** was enhanced by James in 1995 with the introduction of a second boronic acid group to form the diboronic acid sensor **80**.[126,176] This Receptor–Spacer–Fluorophore–Spacer–Receptor system retained the advantage of utilising PET to modulate an "off–on" response to saccharides, while introducing an advanced recognition site. The co-operative action of two boronic acid receptors permitted a number of possible binding modes to occur with saccharides. However, for fluorescence to be restored both boronic acid moieties must be complexed, which requires either an acyclic 2:1 or cyclic 1:1 (saccharide/sensor) complex to form (Figure 20).

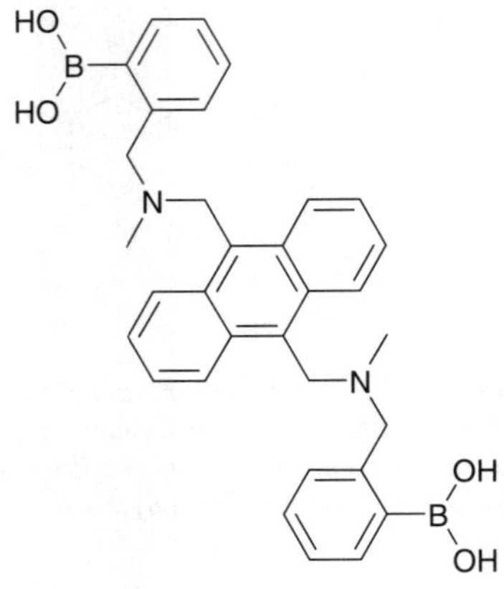

80

Figure 20 *The first rationally designed boronic acid-based fluorescent PET sensor to display selectivity for D-glucose. The receptor–spacer–fluorophore–spacer–receptor assembly requires binding to occur at both receptors in order to restore fluorescence.*

The modification proved successful and fortuitously the spacing of the two boronic acid groups provided an effective binding pocket for D-glucose. D-Glucose complexation occurred with a 1:1 stoichiometry with the saccharide binding to form a macro-cyclic ring. While the inherent selectivity of monoboronic acids is for D-fructose, in this compound the stabilisation derived from the rigid macro-cyclic ring produces a D-glucose selective system. The observed stability constants (K_{obs}) for sensor **80** were 320 M^{-1} with D-fructose, 4000 M^{-1} with D-glucose and 160 M^{-1} with D-galactose in 33 wt% methanol in aqueous buffer at pH 7.77.

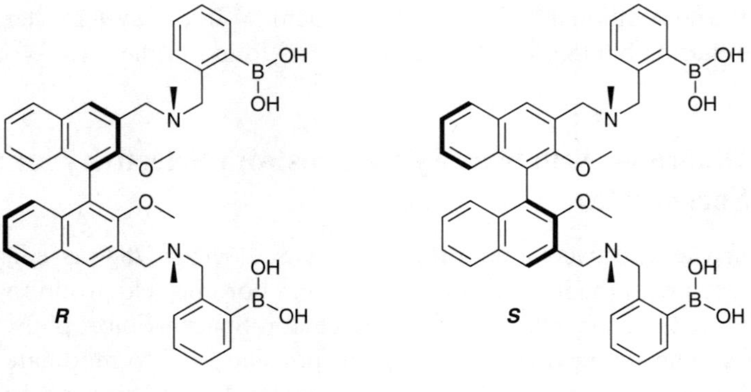

R   *S*

81

Extending the design parameters of sensor **80**, James and co-workers developed the chirally selective sensors *R*- and *S*-**81**.[127,178] Based on the intramolecular fluorescence quenching of 1-1′binaphthyls, sensors *R*- and *S*-**81** demonstrated that selectivity of a diboronic acid appended sensor could be tuned not only towards specific guests, such as monosaccharides, but also towards the specific enantiomers of these guests.

First documented in 1995, the role of sensors *R*- and *S*-**81** as tools for the chiral discrimination of monosaccharides yielded observed stability constants (K_{obs}) of (for sensors *R*-**81** and *S*-**81**, respectively) 10,000 M^{-1} and 5000 M^{-1} with D-fructose, 3200 M^{-1} and 10,000 M^{-1} with L-fructose, 2000 M^{-1} and 2500 M^{-1} with D-glucose and 1300 M^{-1} and 3200 M^{-1} with L-glucose in 33.3 wt% methanol in aqueous buffer at pH 7.77.[127]

Following observations by Houston and Gray[179] that the racemate of sensor **81** bound sugar acids such as tartaric acid with a comparable strength to the monosaccharides [an observed stability constant (K_{obs}) for tartaric acid of 1400 M^{-1} in 33% methanol/water at pH 7.77 (HEPES buffer) was reported] the chiral complexation of these ligands was investigated by Zhao.[178]

82 D-glucaric acid

83 D-glucaric acid

84 D-gluconic acid

85 D-tartaric acid

86 D-sorbitol

It was found that the recognition of D- and L-tartaric acid by sensors *R* and *S*-**81** was strongly pH dependent. At pH 8.3 the fluorescence enhancements behaved as expected. For sensor *R*-**81** with D-tartaric acid and sensor *S*-**81** with L-tartaric acid the fluorescence enhancements were large, conversely for sensor *S*-**81** with D-tartaric acid and sensor *R*-**81** with L-tartaric acid the fluorescence enhancements were small, see Scheme 28.

Accordingly, when the acidity was adjusted to pH 5.6 the fluorescence of sensor *R*-**81** with D-tartaric acid displayed an increase in fluorescence intensity. Quite astonishingly, however, the use of tartaric acid's L-enantiomer with sensor *R*-**81** produced a decrease in the fluorescence intensity. The same relationship was observed with the mirror image host–guest complexes. Sensor *S*-**81** and L-tartaric acid displayed a fluorescence increase, while sensor *S*-**81** and D-tartaric acid displayed a fluorescence decrease. These results signify that sensors *R*- and *S*-**81** have the unusual property of allowing the fluorescence intensity of the reporting signal to be enantioselectively diminished or enhanced, relative to the fluorescence emission of the unbound species, see Scheme 29.

Evaluation of sensors *R*- and *S*-**81** with D-gluconic acid produced a similar enantioselective diminution or enhancement in the fluorescence intensity,

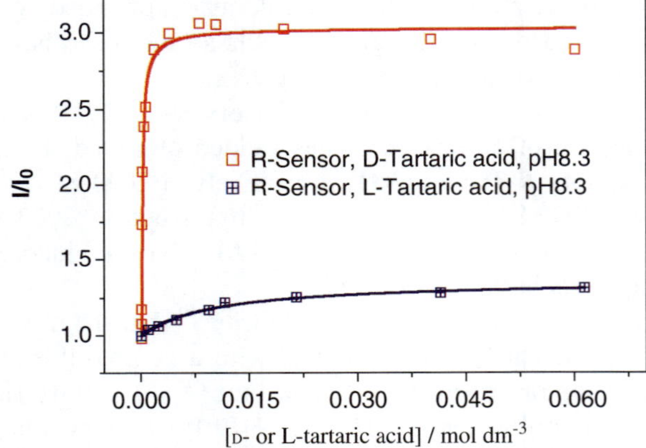

Scheme 28 *D-tartaric acid (plotted in red) and L-tartaric acid (plotted in blue) binding isotherm with sensor R-81 at pH 8.3. [R-81]=5.0 × 10⁻⁶ mol dm⁻³, at pH 8.3 in 0.05 M NaCl (52.1% methanol), [Tartaric acid]=0–0.06 mol dm⁻³, λ_ex at 289 nm, λ_em at 358 nm, 22°C.*

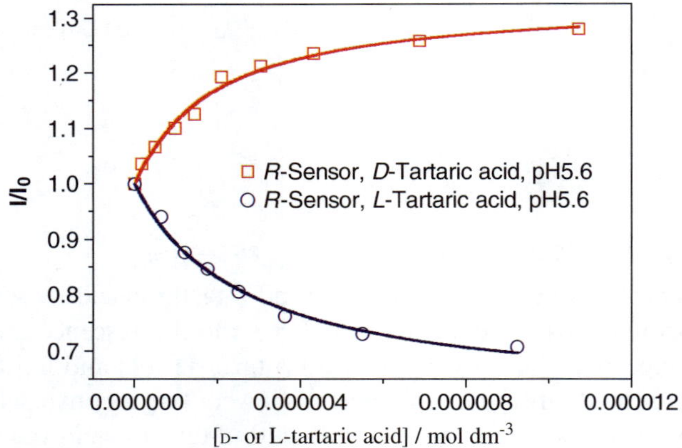

Scheme 29 *D-tartaric acid (plotted in red) and L-tartaric acid (plotted in blue) binding isotherm with sensor R-81 at pH 5.6. [R-81]=5.0 × 10⁻⁶ mol dm⁻³, at pH 5.6 in 0.05 M NaCl (52.1% methanol), λ_ex at 289 nm, λ_em at 358 nm, 22°C.*

curiously this was not the case with D-glucaric acid, D-glucuronic acid or D-sorbitol. It is understood that on complexation of an asymmetric guest to sensors *R-* and *S*-**81**, the orientations of the two receptor units relative to the fluorescent BINOL core are locked in an asymmetrically distorted conformation. These results imply that in all probability the normal PET quenching that mediates fluorescence intensity must be susceptible to the induced-geometrical changes that occur between receptors and fluorophores within individual host–guest complexes; the degree of geometric strain varying on a case-by-case basis.

Considering the effect of geometry on both the stability of the complexes formed and the fluorescence enhancements generated, the two properties appear to be independent. For changes in geometry the observed stability constants (K_{obs}) increased up to ~25-fold, while fluorescence intensity increased only up to ~3 fold. The geometry of the receptor therefore has a far greater influence on substrate recognition than it does on fluorescence.

The observed stability constants (K_{obs}) for (sensors *R*-**81** and *S*-**81**, respectively) were 520,000 M^{-1} and 260,000 M^{-1} with D-tartaric acid, 390,000 M^{-1} and 280,000 M^{-1} with L-tartaric acid in 52.1 wt% methanol in water at pH 5.6.

A number of diboronic acid fluorescent sensors have been synthesised with significantly modified structures. The interboronic acid distance, responsible for defining the dimensions of the binding cleft proved particularly important. Where a reduced distance between the boronic acids was chosen by James with sensor **87**,[180] the sensor was found to differentiate between monosaccharides based on the number and proximity of the diol groups the saccharide contained. Ligands such as propane-1,3-diol, with only one binding site, and D-glucose, where the two diol groups were separated by one or more carbons, were entirely excluded from the binding pocket, whereas smaller monosaccharides with dual adjacent diol groups such as D-sorbitol, complexed strongly. The observed stability constants (K_{obs}) for sensor **87** were 350 M^{-1} with D-sorbitol and 330 M^{-1} with D-fructose in 300:1 (v/v) water/methanol at pH 8.0, while the fluorescence changes for D-glucose and propane-1,3-diol were so small that observed stability constants (K_{obs}) could not be determined (Figure 21).

Conversely Linnane found that a gross increase in the interboronic acid distance, with a rigid spacer such as with sensor **88**,[181] displayed a significant loss in selectivity and sensitivity. The observed stability constants (K_{obs}) for sensor **88** were 170 M^{-1} with D-fructose, 91 M^{-1} with D-glucose and 72 M^{-1} with D-galactose in 33 wt% methanol in aqueous buffer at pH 7.77.

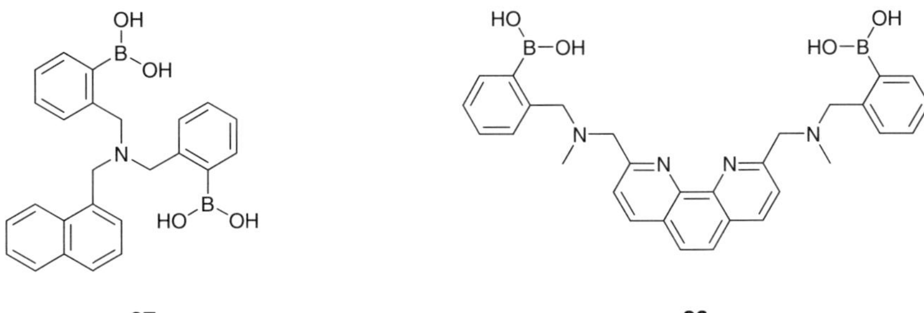

87　　　　　　　　　　**88**

Figure 21 *Diboronic acid-based fluorescent PET sensors with varied interboronic acid distance. Sensor **87** demonstrated enhanced selectivity for D-sorbitol, while sensor **88** displayed neither selectivity nor sensitivity towards monosaccharides.*

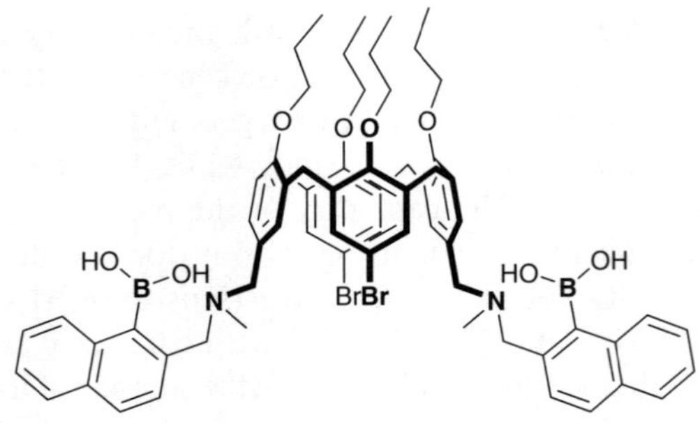

89

In an effort to develop sensors with dual boronic acid receptors favourably located so as to enhance D-glucose selectivity, boronic acid receptor groups were mounted onto a calixerane scaffold by Linnane to generate sensor **89**.[182] This elegant approach had the potential to allow a diverse range of synthetic functionality to be designed into the sensors, unfortunately the sensor displayed a poor affinity for monosaccharides. The observed stability constants (K_{obs}) were 115 M^{-1} for D-fructose (forming an acyclic 2:1 saccharide/sensor complex) and 24 M^{-1} for D-glucose (forming a cyclic 1:1 saccharide/sensor complex) in 33 wt% methanol in water at neutral pH.

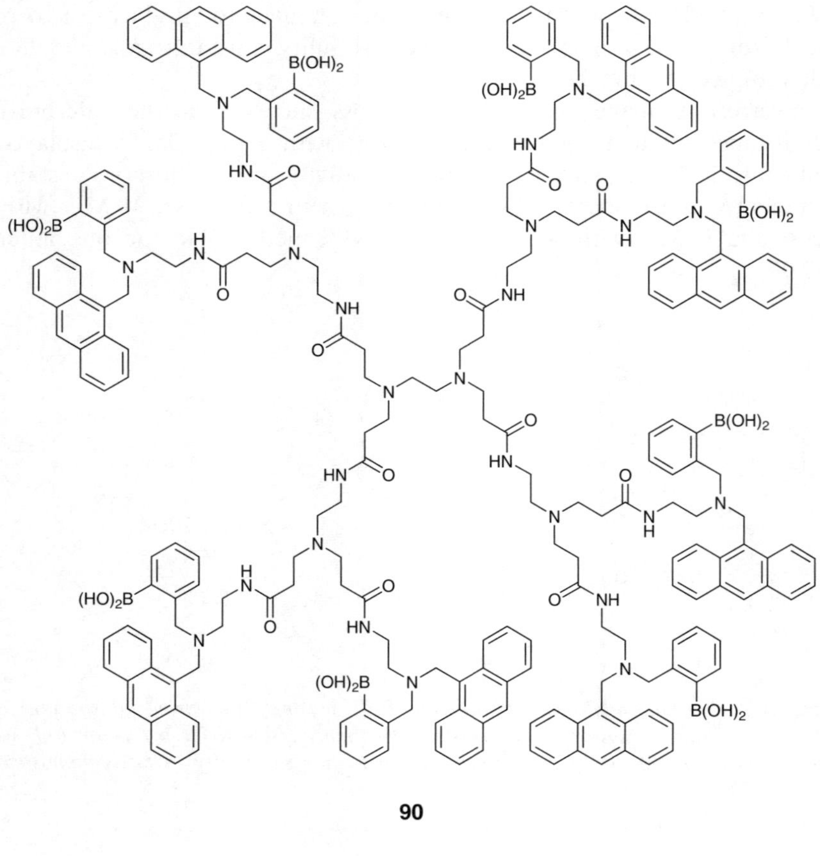

90

The boronic acid-based starburst dendrimer **90** investigated by James displayed an enhanced sensitivity for all of the monosaccharides it was examined against.[183] In essence, the dendrimer functioned as a saccharide "sponge". While the dendritic boronic acid is highly flexible and one could therefore expect little pre-organisation within the binding pocket, the boronic acid–saccharide complex was surprisingly stable. To determine the binding motifs used within this compound, a number of deoxy-, deoxymethyl- and partially protected saccharide derivatives were examined. The results supported the view that even within such a flexible sensor the co-operative action of the boronic acids to form 2:1 (boronic acid group/saccharide) complexes led to an enhanced binding ability. Moreover, the results indicated that binding to D-galactose must occur primarily through the hydroxyls on the 1 and 2 positions. This is apparent from the observed stability constants (K_{obs}) reported for dendrimer **90**, which were 27,000 M^{-1} with D-galactose, 16,900 M^{-1} with D-fructose and 740 M^{-1} with D-glucose in pure methanol.[183]

91

When the hydroxyl group on the anomeric centre was substituted for a methoxy group, as in 1-methyl-α-D-galactopyranoside **91**, the observed stability constant (K_{obs}) diminished to 40 M^{-1}, a 675-fold reduction in the observed stability constant (K_{obs}).

92

Kukrer found that on addition of saccharide the diboronic acid squaraine **92** produced a fluorescence response with an emission maxima at 645 nm and a shoulder at 695 nm.[184] This emission occurs at the blue end of the near-infrared optical window (600–1300 nm) for human skin.[185]

The biomedical interest in the near-infrared section of the electromagnetic spectrum is derived from the comparatively high absorption and scattering of light by living tissues observed outside of these wavelengths.[186] Moreover, signal reduction is coupled with increased signal to noise ratios generated from the autofluorescence of mammalian vital organs and bodily fluids when subject to wavelengths below ~600 nm.[187] While interference from these properties can be readily overcome in routine analysis, the development of fluorescent sensors with excitation and emission wavelengths inside the near-infrared band permits the biomedical imaging of sensors within bulk tissue as well as the potential transdermal monitoring of analytes *in vivo*.[187]

D-Fructose produced a 25% increase in fluorescence intensity on addition to the sensor in an ethanol/water (50 mM carbonate buffer) solution at pH 10. Unfortunately, D-glucose, D-galactose and D-mannose only produced an observed change in the fluorescence intensity of 8%.

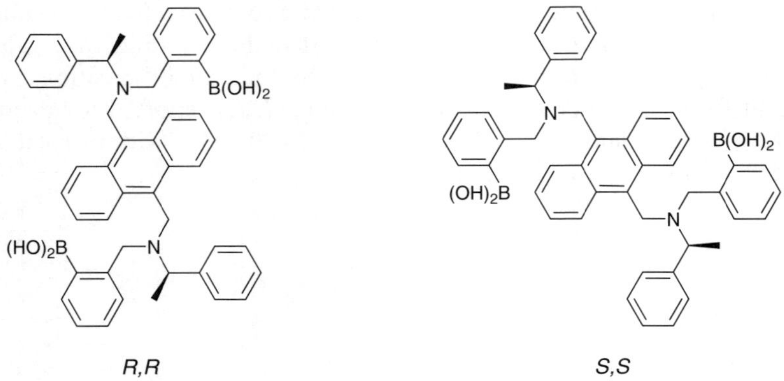

R,R   S,S

93

The design assembly employed in these sensory systems appears to have come full circle. Zhao recently reported chiral boronic acid-based sensors R,R- and S,S-**93**, which are direct structural derivatives of the original D-glucose selective sensor **80**.[188,189]

The pH titrations demonstrated that sensors R,R- and S,S-**93** were highly selective for sugar acids such as tartaric, glucaric, gluconic and glucuronic acid; with chemoselectivity of up to 11,000:1 (pH 5.6) being reported between species and enantioselectivity between the D- and L- forms of the ligands of up to 500:1 (pH 7.0) being reported with sensors R,R- and S,S-**93**. The same titrations were performed with monoboronic acid analogues, these analogues displayed no enantioselective discrimination between D- and L- tartaric acid indicating that for enantioselectivity a 1:1 cyclic complex must be formed.

The observed stability constants (K_{obs}) for (R,R-**93** and S,S-**93**, respectively) were 830,000 M^{-1} and 10,000 M^{-1} with D-tartaric acid, 8500 M^{-1} and 580,000 M^{-1} with L-tartaric acid in 52.1 wt% methanol in water at pH 5.6.

4.7   Ditopic Sensors

Ditopic sensors take advantage of two non-equivalent recognition sites. If one is a boronic acid unit, selectivity can be introduced for saccharidic species such as uronic acids, amino sugars and so forth. To achieve this, the second recognition unit must display, and be positioned, such that effective complementarity is observed for both types of functionality present on the guest molecule.

The fluorescent PET sensors **94** and **95** prepared by Cooper demonstrated selectivity for an ammonium terminus at the monoazacrown ether receptor and selectivity for saccharides at the boronic acid receptor. This dual-targeted approach induced D-glucosamine hydrochloride selectivity within the system.[190,191]

D-glucosamine
hydrochlorid

94

95

Curiously, the azacrown ether imparts little if any additional stability to the complex on binding. Nevertheless, binding at the azacrown ether is a pre-requisite of generating a fluorescent response as it directly participates in PET, the system requiring both recognition units to complex for fluorescence to be restored. The observed stability constants (K_{obs}) for sensors **94** and **95** with D-glucosamine were 18 M^{-1} and 17 M^{-1}, respectively, in 33.2 wt% ethanol/water at pH 7.18 (triethanolamine buffer).

Boronic acid-based fluorescent PET sensors developed for the recognition of simple monosaccharides have been extended to include ditopic recognition sites and so introduce selectivity for a diverse range of guest species.

96

Developed around the D-glucose sensor **80** the allosteric diboronic acid biscrown ether **96** prepared by James displayed a fluorescence increase upon binding D-glucose.[192] Addition of metal cations with a similar ionic diameter to potassium favour a 1:1 (metal ion/sensor) binding motif, causing the 15-crown-5-ether receptors to sandwich the metal ion. The metal-induced conformational change disrupts the 1:1 (saccharide: sensor) complex with D-glucose, restoring PET and thus quenching fluorescence.

The observed stability constants (K_{obs}), (and ionic diameters) for the metal cations were 19 M^{-1} (2.04 Å) with Na^+, 35 M^{-1} (2.36 Å) with Sr^{2+}, 63 M^{-1} (2.76 Å) with K^+ and 2500 M^{-1} (2.70 Å) with Ba^{2+} in 33.3 wt% methanol in water in the presence of 0.03 mol dm^{-3} D-glucose. The fluorescence changes for Li^+ (1.52 Å) and Cs^+ (3.40 Å) were so small that observed stability constants (K_{obs}) could not be determined.

97

Designed by Shinkai and co-workers **97** also takes advantage of two distinct recognition moieties.[193,194] The monoboronic acid receptor allows selectivity towards saccharides to be introduced to the sensor and modulated through PET. When Zn(II) is co-ordinated with the two phenanthroline nitrogen atoms, the selectivity for saccharides remains unchanged. However, the interaction of the metal centre with carboxylate groups on guest species significantly enhances the selectivity for sialic and uronic acids. In keeping with the expected trends for monoboronic acids, the observed stability constants (K_{obs}) for sensor **97** were 79 M^{-1} for D-galacturonic acid, 50 M^{-1} for D-galactose and 320 M^{-1} for D-fructose. With the introduction of Zn(II), the values increased to 2510 M^{-1} for D-glucuronic acid, 1260 M^{-1} for D-galacturonic acid, 200 M^{-1} for sialic acid, while remaining nearly constant for the monosaccharides, 20 M^{-1} for D-galactose and 250 M^{-1} for D-fructose.

98

The boronic acid, guanidinium appended fluorescent PET sensor **98** developed by Wang and co-workers displayed selectivity for D-glucarate.[195] As with other ditopic sensors both receptor units must be occupied for fluorescence to be restored. In this instance, the boronic acid participates in diol complexation and the guanidinium participates in carboxylate complexation.

The observed stability constants (K_{obs}) for sensor **98** were 5140 M^{-1} with D-glucarate, 1450 M^{-1} with D-gluconate, 1300 M^{-1} with D-sorbitol, 62 M^{-1} with D-glucose and 46 M^{-1} with D-glucuronic acid in 50% methanol in 0.1 M aqueous HEPES buffer at pH 7.4.

99

Deetz and Smith have prepared a heteroditopic ruthenium(II) bipyridyl receptor with both saccharide and phosphate-binding sites **99**[196] The receptor displays enhanced affinity for phosphorylated saccharides over normal saccharides. The system is also positively allosteric for saccharides in the presence of phosphate – the association constants increase by two orders of magnitude when titrations with saccharide are conducted in sodium phosphate buffer. The observed stability constants (K_{obs}) for **99** were 32 M^{-1} for D-fructose, 1580 M^{-1} for D-fructose in 10 mM phosphate buffer and 1260 M^{-1} for D-fructose-6-phosphate in water at pH 7.3.

100   **101**

The ditopic fluorescent sensors **100** and **101** have been developed by Koskela as reversible AND logic gates with selectivity for potassium fluoride.[197] The binding of fluoride with boronic acids is well documented.[198,199] In this case the sp^2 hybridised boronic acid, which is a hard Lewis acid, interacts strongly with the fluoride anion, which is a hard Lewis base, to become sp^3 hybridised. The potassium cation is held *in situ* partly by the crown ether and partly by the electrostatic interaction with the fluoride anion. This co-operative complexation allows the cationic and anionic guests to be bound to the host as an ion pair, while allowing the host to discriminate between potassium fluoride and other similar ion pairs such as potassium chloride and potassium bromide (Scheme 30).

The dynamic characteristics of this sensor are particularly attractive. Not only are both guests required at the binding sites to generate a fluorescence response, but it is possible to add and remove individual guests producing a controlled and reversible read-out signal derived from the AND logic functionality inherent to the sensor. This example is illustrated in Scheme 30 and provides an elegant representation of the dynamic and versatile disposition of these recognition systems, permitting selective and reversible binding to be designed into sensors and switches alike.

The observed stability constants (K_{obs}) for sensor **100** were 10,000 M^{-1} with potassium and 320 M^{-1} with fluoride. Those for sensor **101** were 50,000 M^{-1} with potassium and 250 M^{-1} with fluoride in methanol.

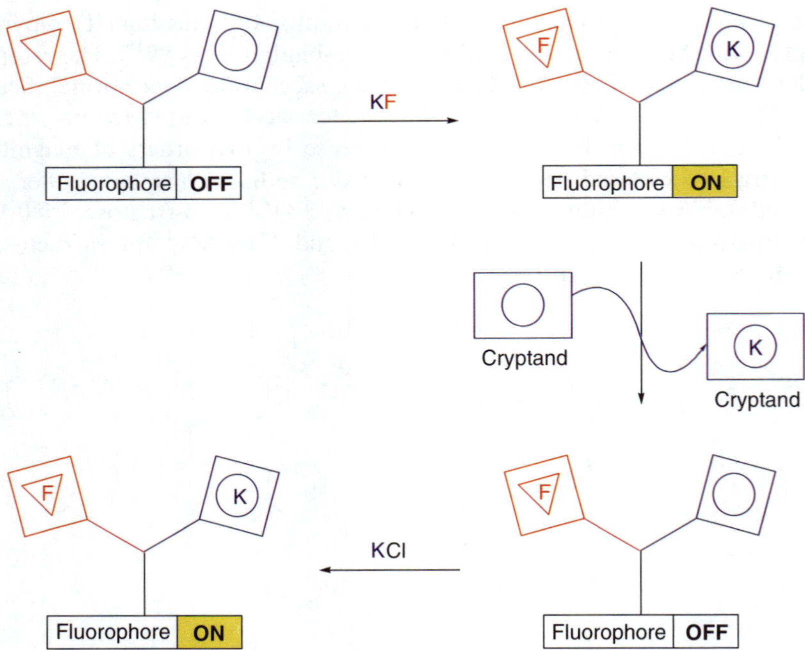

Scheme 30 *Schematic representation of the ditopic fluorescent PET sensors **100** and **101**, demonstrating selective and reversible binding of potassium fluoride with concurrent modulation of fluorescence intensity. The initial unbound complex is "off", addition of KF generates a large increase in fluorescence intensity turning the system "on", removal of potassium by addition of [2.2.2]-crypt-and (4,7,13,16,21,24-hexaoxa-1,10-diazabicyclo[8.8.8]-hexacosane) quenches fluorescence returning the sensor to an "off" state, subsequent addition of potassium chloride re-introduces potassium to the system and restores fluorescence to the "on" state.*

4.8   Other Fluorescent Sensors

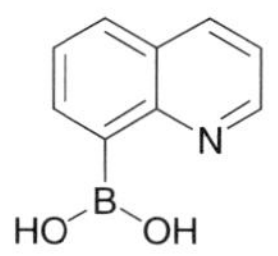

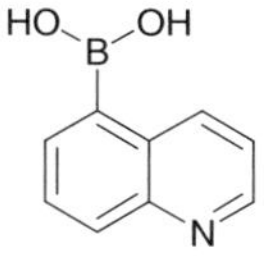

102

103

Wang has recently reported that 8-quinoline boronic acid **102** responds to the binding of D-fructose with over 40 fold increases in fluorescence intensity, while D-glucose only produced a 3-fold fluorescence increase.[200] The authors ascribe the fluorescence changes to environmental factors and not a change in the hybridisation of the boron. The observed stability constants (K_{obs}) for **102** were 108 M^{-1} for D-fructose and 7.5 M^{-1} for D-galactose in water at pH 7.4 (phosphate buffer). The 5-quinoline boronic acid **103** was also investigated and enhanced binding towards saccharides was observed.[201] The observed stability constants (K_{obs}) for **103** were 924 M^{-1} for D-fructose and 90 M^{-1} for D-galactose in water at pH 7.4 (phosphate buffer).

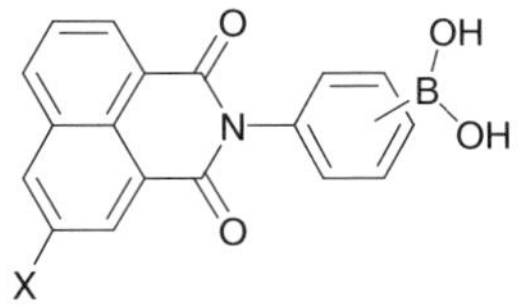

104: X=H, meta
105: X=H, ortho
106: X=NO$_2$, meta
107: X=NH$_2$, meta

Heagy and Lakowicz have been investigating *N*-phenylboronic acid derivatives of 1,8-naphthalimide **104** and **105**, and with these systems the fluorescence is substantially quenched (*ca.* 5-fold) on saccharide binding[202,203] The fluorescence change has been ascribed to PET from the boronate to the naphthalimide fluorophore. The nitro-derivative **106** was particularly interesting because it displayed dual fluorescence and was particularly sensitive for D-glucose.[204] Interestingly, sensor **107** shows selectivity for D-galactose.[205] The observed stability constants (K_{obs}) for **104**, **105** and **106** were 769, 625 and 476 M^{-1} for D-fructose and 21.7, 17.5 and 38.5 M^{-1} for D-glucose in water at pH 7.5, 7.5 and 8.0, respectively (phosphate buffer).

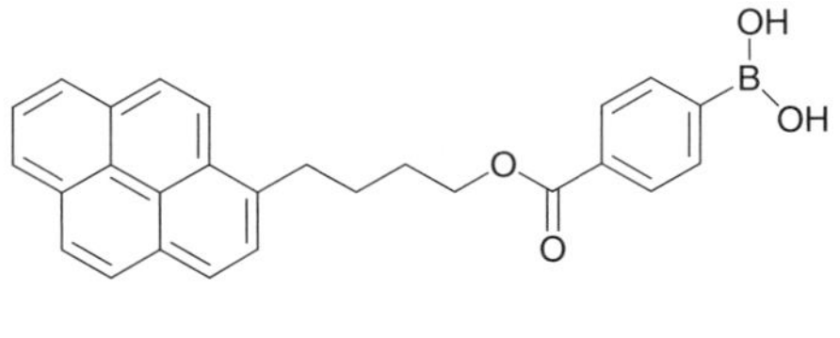

108

Hayashita and Teramae have prepared an interesting fluorescent ensemble comprising of compound **108** and β cyclodextrin[206] The system displays fluorescence enhancement on saccharide binding and as expected for a monoboronic acid the highest binding was observed with D-fructose. The observed stability constants (K_{obs}) for **108** were 2515 M^{-1} for D-fructose and 79 M^{-1} for D-glucose in 2% (v/v) DMSO/water at pH 7.5.

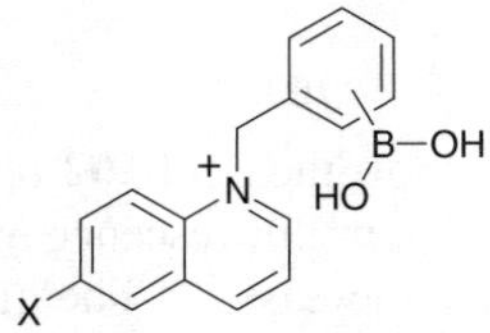

109: X=OCH$_3$, meta
110: X=OCH$_3$, ortho
111: X=OCH$_3$, para
112: X=CH$_3$, meta
113: X=CH$_3$, meta
114: X=CH$_3$, para

Lakowicz and Geddes have explored the use of several quaternized quinolium boronic acids **109**, **110**, **111**, **112**, **113** and **114** for D-glucose monitoring within contact lenses.[207–209]

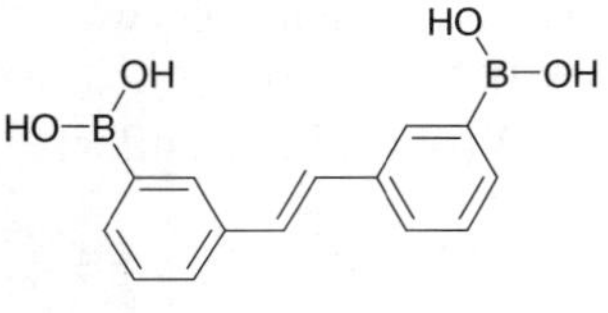

115

Diboronic acids can bind monosaccharides selectively, where the 1:1 binding creates a rigid molecular complex.[112,210–213] This rigidification effect can also be utilised in designing fluorescent sensors for disaccharides. Sandanayake has investigated the binding of diboronic acid **115** with disaccharides in basic aqueous media.[214] It is known that excited stilbene is quenched by radiationless decay *via* rotation of the ethylene double bond. Obstruction of this rotation leads to increased fluorescence.[215] The rigidification of **115** by disaccharide binding increases the stilbene fluorescence. In particular, the disaccharide D-melibiose shows higher selectivity for **115** than other common disaccharides.

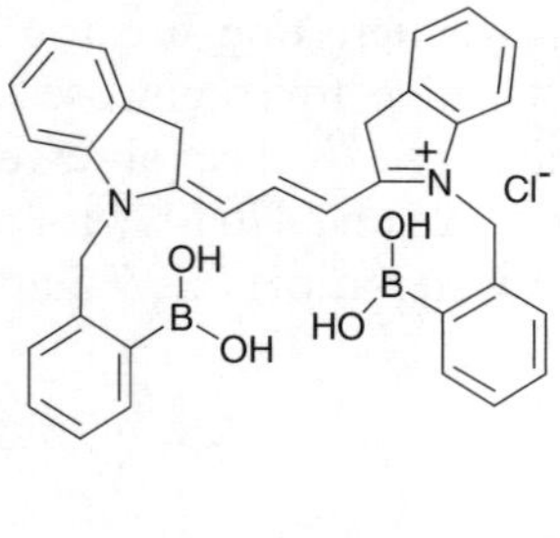

116 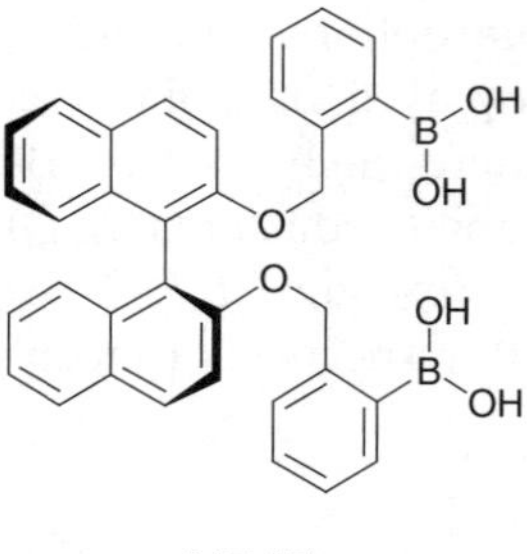   **117 (***R***)**

Takeuchi has used molecular rigidification of cyanine diboronic acid **116** to generate a fluorescence increase with added saccharides.[216] The observed stability constants (K_{obs}) for **116** were 130,000 M^{-1} for D-fructose and 1400 M^{-1} for D-glucose in 1:1 (v/v) methanol/water at pH 10.0 (carbonate buffer).

Rigidification has also been used with a diboronic acid appended binaphthyl **117(R)** to develop a chirally discriminating system.[217] The observed stability constants (K_{obs}) for **117(R)** were 8600 M^{-1} for D-fructose, 14,400 M^{-1} for L-fructose, 2100 M^{-1} for D-glucose and 1900 M^{-1} for L-glucose in 1% (v/v) methanol/water at pH 10.8, respectively (carbonate buffer).

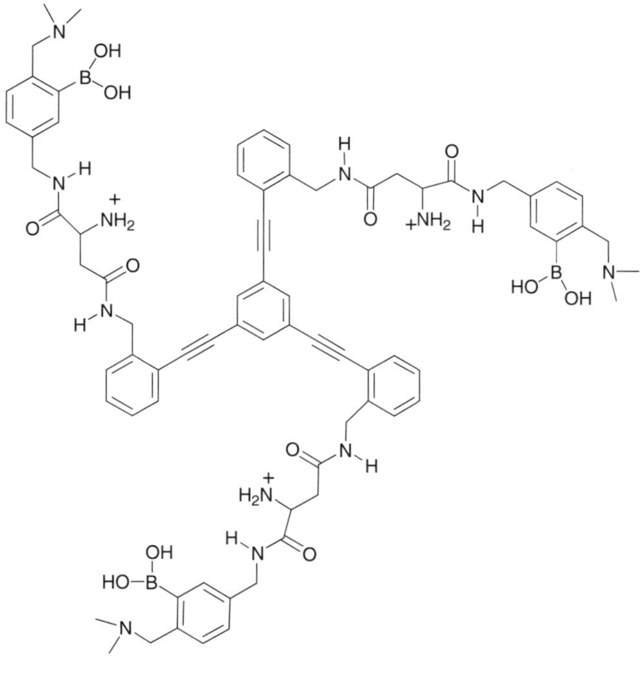

118

Anslyn has prepared a heparin-selective sensor **118**.[218] Interaction with unfractionated heparin (UFH) and low-molecular-weight heparin (LMWH) causes conformational restriction of the "arms" of the receptor and a reduction in fluorescence intensity. The observed stability constant (K_{obs}) for **118** was 1.4 $\times$ 10^8 M^{-1} for UFH in water at pH 7.4 (HEPES buffer).

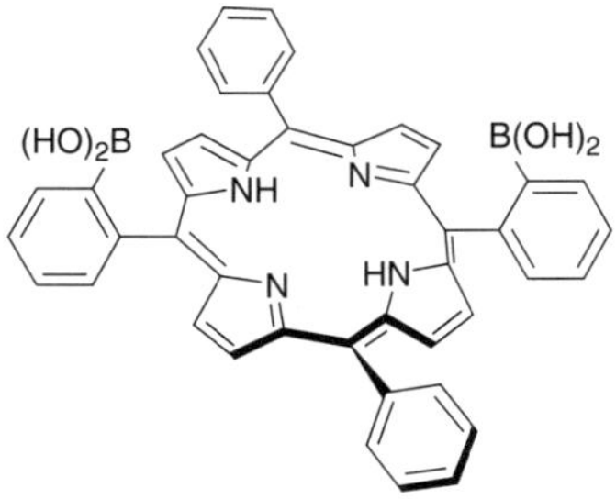

119

Kijima has developed a D-lactulose selective system **119** based on a diboronic acid porphyrin.[219] The spatial disposition of the two boronic acids in **119** produces the perfect "cleft" for the disaccharide D-lactulose. The observed stability constant (K_{obs}) for **119** was 560 M^{-1} for D-lactulose in methanol. Other disaccharides produced no spectral changes.

120

Norrild has developed an interesting diboronic acid system **120**.[220] This system works by reducing the quenching ability of the pyridine groups of **120** on saccharide binding. The structure of the complex was determined to be to a 1,2:3,5 bound α-D-glucofuranose. Evidence for the furanose structure was obtained from ^{1}H- and ^{13}C-NMR data with emphasis on the information from $^1J_{C-C}$ coupling constants. The observed stability constant (K_{obs}) for **120** was 2500 M^{-1} for D-glucose in water at pH 7.4 (phosphate buffer). D-Fructose and D-galactose were bound much weaker than D-glucose and the binding constants could not be determined using a simple 1:1 binding model.

121: *ortho*
122: *meta*
123: *para*

Mohr has prepared three isomeric hemicyanine dye molecules which show fluorescent enhancement at 600 nm on the addition of saccharides.[221] The observed stability constants (K_{obs}) for **121, 122** and **123** were 280, 40 and 200 M^{-1} for D-fructose and 4, 15 and 14 M^{-1} for D-glucose in water at pH 7.3 (phosphate buffer).

124

A novel approach has been taken by Hamachi who has coupled a natural receptor protein Concanavalin A (ConA) with a fluorescent boronic acid PET system to prepare a semi-synthetic biosensor **124**.[222] The system illustrates how both synthetic and natural systems can be combined to generate biosensors with enhanced selectivity towards specific oligosaccharides. The observed stability constants (K_{obs}) for **124** (R) were 708 M^{-1} for D-fructose and 151,000 M^{-1} for D-palatinose in water at pH 7.5 (HEPES buffer).

125

An interesting system **125** has been developed by Zhang and Zhu, which contains an electron rich tetrathiafulvalene (TTF) unit. It is believed the TTF plays a similar role to the nitrogen in **78**. In the absence of saccharides, the fluorescence of the anthracene is quenched by PET from the TTF unit. However, when saccharides are added the boronate ester formed is a better Lewis acid (electron acceptor) and PET to the boronate dominates resulting in an increase in the anthracene fluorescence.[223]

4.9 Amine–Boron (N–B) Interactions

The so-called coordinative or dative, nitrogen to boron (N–B or alternatively N→B) bonds have been studied for more than 100 years. The strength of N–B bonds depend greatly on the substituents at both atoms; electron-withdrawing groups increase the Lewis acidity of boron, while electron-donating groups increase the Lewis basicity at nitrogen. In considering the bond strength, it is necessary to balance these electronic factors against the counteracting steric requirements of these same substituents. An investigation of 144 compounds with coordinative N–B bonds concluded that steric interactions as well as ring strain (in the case of cyclic diesters) weaken and elongate the N–B bond, which occurs with a concurrent reduction in the tetrahedral geometry of the boron centre.[224]

The *N*-methyl-*o*-(phenylboronic acid)-*N*-benzylamine **126** system has been investigated separately by Wulff, Anslyn and within the T. D. James research group.[72,174,225] Scheme 31 depicts a general model where, at one extreme, the acyclic forms (**126** and **127**) illustrate no N–B interaction and at the other, the cyclic forms (**129** and **130**) illustrate a full N–B interaction; the species existing in equilibrium. Species **128** involves a protonated nitrogen, therefore the ammonium cation precludes any N–B interaction.

The energy of the N–B interaction has been calculated from the stepwise formation constants of potentiometric titrations. Based on the relative stabilities of ternary phosphate complexes, it was calculated that the upper and lower limits of the N–B interaction must be bound between ~ 15 and 25 kJ mol^{-1} in *N*-methyl-*o*-(phenylboronic acid)-*N*-benzylamine.[72] This value is in good agreement with computational data, which estimated the N–B interaction to be 13 kJ mol^{-1} or less in the absence of solvent.[226] To qualify this in terms of familiar

126

127

$pK_a = 5.3$

$pK_a = 12.07$

128

129

130

Scheme 31 *The extent of the interaction between nitrogen and boron is illustrated within the upper and lower bounds of possible contact depicted as the cyclic and acyclic forms.[72]*

bonding motifs the energy of the N–B interaction is about the same as that of a hydrogen bond.

As we have discussed, the strength of this interaction is a central feature in many fluorescent PET sensors, where the N–B interaction plays a pivotal role in signalling the binding event. If the interaction were much stronger, then the binding of a diol would not be able to disrupt the N–B interaction sufficiently so as to modulate a change in fluorescence. By the same token, if the interaction were much weaker, then there would be no significant intramolecular N–B interaction to disrupt in the first place.

For some time, the formation of a direct bond between nitrogen and boron was assumed to be responsible for the fluorescence enhancement seen when boronic acids bound diols. This interpretation does, however, raise certain questions.

The fluorescence revival in these systems functions as a digital "off–on" response. Fluorescence emission returns to the same maximal value regardless of the observed stability constant (K_{obs}) of the ligand or the pK_a' of the resulting boronate ester. It is known that the acidity of boron in other systems directly influences the strength of N–B bond. Therefore, if a direct N–B bond did modulate PET, we might necessarily expect the fluorescence response to vary as a function of the degree of acidity or strength of complexation, this is not the case.[227] Moreover, the numerical values do not agree with the interpretation of a direct N–B bond.[72]

The recent evaluation of sensor **93**, the first fluorescent diboronic acid sensor to be reported as a single-crystal X-ray structure both in its bound and unbound state, has recently been published.[188] In the case of the unbound receptor, the geometry at boron is trigonal planar. This is important as the absence of deviation from planarity implies that there is no direct N–B Lewis base–Lewis acid bond at boron.

When bound to tartaric acid, the complex was crystallised from a methanol and dichloromethane solution. In the tartrate complex, two molecules of methanol, one at each boron centre are bound through their oxygen atoms to their respective boron centres. While the hydrogen atoms of methanol were not directly located, it is not unreasonable to infer from the geometry that each oxygen atom will concurrently bind to the boron centre and hydrogen bond to the adjacent nitrogen atom. Oxygen to nitrogen bond distances of around 2.7 Å were reported in this case, Figure 22.

While speculative, this structural interpretation of the interaction between boronic acid and the proximal tertiary amine through a bound protic solvent molecule (solvent insertion into the N–B bond) corresponds well with contemporary computational and potentiometric titration data, in which the formation of intramolecular seven-membered rings should not be ignored.[67,72,228,229] The values for the bond length (from the X-ray crystal structure) and for the bond strength (from the potentiometric titrations) are those that would be expected for a hydrogen-bonding interaction manifested through a bound solvent molecule at the boron centre.

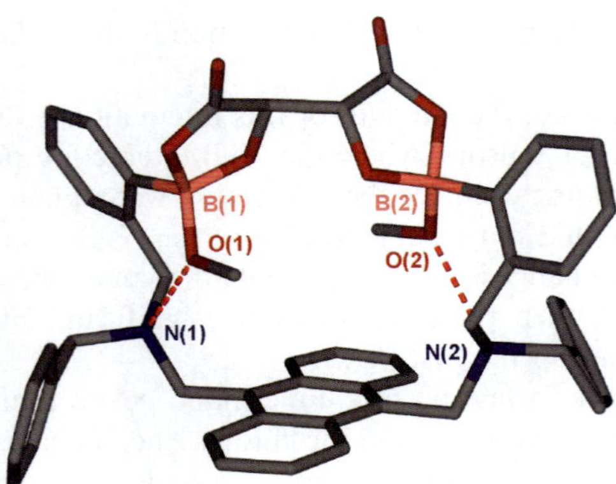

Figure 22 *The single-crystal X-ray structure of the S,S-diboronic acid (S,S-**93**)–L-tartaric acid complex isolated by James and co-workers.[188] While the hydrogen atoms of methanol were not directly located it can be inferred that the geometry between the bound and unbound receptors will be similar, each oxygen atom will therefore concurrently bind to the boron centre and hydrogen bond to the nitrogen atom. Oxygen to nitrogen bond distances of around 2.7 Å were reported [N(1) . . . O(1)=2.655 Å and N(2) . . . O(2)=2.693 Å]. Atoms marked in red represent oxygen, pink boron, grey carbon and blue nitrogen. For clarity, hydrogen atoms are not displayed. The red-dotted lines represent hydrogen bonds.*

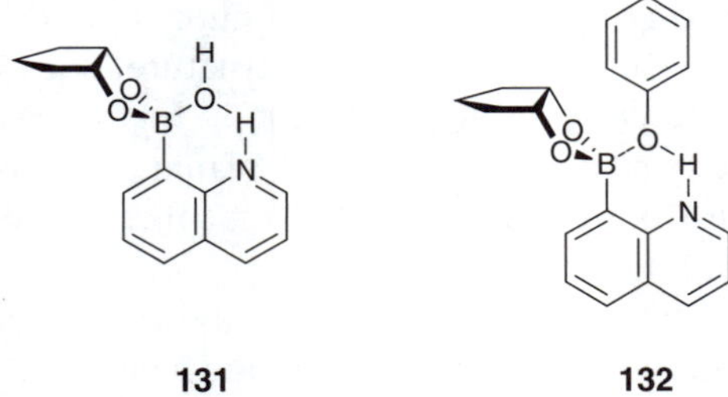

131   **132**

Figure 23 *Morrison's proposed complexes of the cis-1,2-cyclopentanediol ester of 8-quinolineboronic acid with solvent water and phenol molecules bridging the nitrogen and boron centres.[230] Blue bonds are used to represent dative bonding.*

The idea postulated is by no means a new one. An infrared study into the interaction between nitrogen and boron in a similar system came to a similar "tentative conclusion" in 1964.[230] The experimental rationale was based on comparing two emergent peaks in infrared spectra to similar peaks in known model systems. The results indicated that in carbon tetrachloride the interaction between the nitrogen and boron of 8-quinolineboronic acid could be modulated by either water or phenol bound to the boron centre at oxygen (Figure 23).

Anslyn recently performed a detailed structural investigation of the N–B interaction in *o*-(*N,N*-dialkyl aminomethyl) arylboronate systems.[231] From detailed ^{11}B-NMR measurements (and X-ray data), it was shown that in an aprotic solvent, the N–B dative bond is usually present. However, in a protic media, solvent insertion of the N–B, occurs to afford a hydrogen-bonded zwitterionic species.

So, what is the take-home message about the N–B interaction (in these systems). Until recently, we would have been reluctant to make any sweeping staements but thanks to the seminal publication from Anslyn[231] and current work, from a number of other groups.[72,226,227,231] The N–B interaction can be ascribed to a hydrogen-bonding interaction mediated through a bound solvent molecule. In other words, the N–B interaction in protic media such as water or methanol should not be represented as **129** but, rather the solvent inserted form **133**.

From a design perspective, even though the mechanism responsible for conferring binding information from boron to nitrogen is indirect, what can be said with certainty is that this recognition unit is an effective one in saccharide recognition and should continue to be incorporated into sensory systems.

4.10 The Importance of Pyranose to Furanose Interconversion

4.10.1 Pyranose to Furanose Interconversion as a Function of Time and Water

In refining the selectivity of boronic acid sensors for saccharides, the structure of the guest species must be addressed. Although there is a long history of research into the structural character of boronic acid–saccharide complexes, the rapid isomerisation of monosaccharides in water precludes the description of a simple generic binding motif. As we have mentioned earlier the hemiacetal ring of a monosaccharide is readily cleaved in water, often reforming rings of different sizes and anomeric configurations. The equilibrium between linear, pyranose and furanose configurations as well as the α and β anomers of the pyranose and furanose rings, substantially increases the number of possible structures that may be formed on complexation.

Sensor **80** provided the first structural elucidation of a diboronic acid sensor with D-glucose complexed within the binding site. The ^{1}H-NMR spectrum of this complex indicated that in deuterated methanol the D-glucose was bound in the α-pyranose form at the 1,2 and 4,6 positions, as in Figure 24.[176]

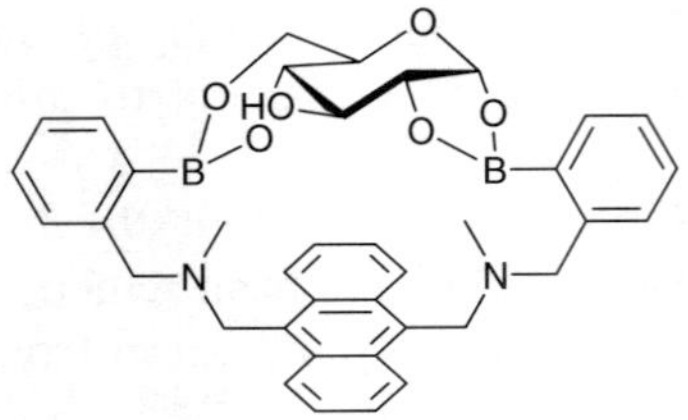

Figure 24 *The initial 1,2:4,6 complex formed between sensor* **80** *and* D-*glucose in MeOD.[176]*

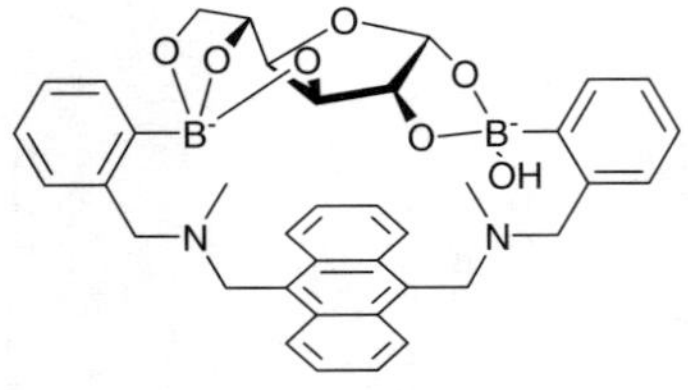

Figure 25 *The thermodynamically stable 1,2:3,5,6 complex observed to form between sensor* **80** *and* D-*glucose in basic aqueous media (given the current understanding of N–B interactions the boron bound at the 1,2 position is illustrated as a tetrahedral boronate).[232]*

A re-examination of this work was conducted by Norrild and Eggert.[232] Employing ^{13}C-1 and ^{13}C-6 labelled D-glucose the $^{1}J_{C1-C2}$ and $^{1}J_{C5-C6}$ coupling constants were monitored. Exploiting the observed reduction in the $^{1}J_{CC}$ value upon formation of a five-membered boronic ester,[233] the analysis determined that the previous ^{1}H-NMR assignment was correct, but that the interpretation was only valid as the initial complex formed under anhydrous conditions. With time the α-D-glucopyranose isomerised to the α-D-glucofuranose form. In deuterated methanol this process was slow, 20 h elapsed before the emergence of new NMR peaks became clear, with a complete disappearance of the original α-D-glucopyranose signals occurring after eight days. However, if water was introduced to the system, isomerisation was accelerated dramatically and after 10 min in a 1:2 water/methanol solution isomerisation had gone to completion.

In the case of sensor **80**, it was concluded that once formed the complex in Figure 24 re-arranges to the thermodynamically more stable 1,2:3,5,6 bound α-D-glucofuranose complex, Figure 25, as a function of time and the water content of the medium. This binding pattern is comparable with that of other known structures, such as the commercially available diacetone-D-glucose **134**.

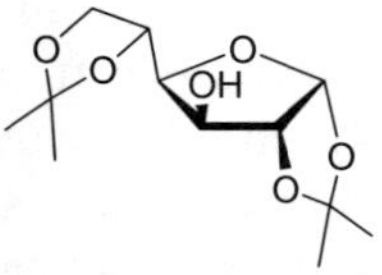

134

Scheme 32 *The conformationally locked C1,C2-synperiplanar diol group of the α-D-glucofuranose ring provides an ideal binding site for boronic acids.*[234]

Indeed, the furanose form of the ring is the one that might be intuitively expected. For an O–B–O fragment to bridge the gap between conformationally locked synclinal (axial and equatorial) hydroxyl groups on a pyranose ring, a substantial degree of torsional strain would be induced across the newly formed cyclic diester. In an effort to understand the significance of this torsional strain, Nicholls and Paul undertook the molecular modelling of a series of boronic acid–cyclic saccharide complexes to determine the structure of the thermodynamically stable boronate esters formed.[234] It was found that the vicinal diols involved in complexation must posses, or at least be able to adopt, a *syn*-periplanar arrangement.

Anti-periplanar hydroxyl groups therefore fail to form boronic acid–diol complexes. Synclinal hydroxyl groups can adopt *syn*-periplanar geometry in instances when the energy lost due to induced distortions in the saccharide ring are outweighed by the energetic stability gained from complexation. But generally the most stable boronic acid–diol complexes occur between boronic acids and the conformationally locked *syn*-periplanar hydroxyl groups on furanose rings (Scheme 32).[234]

This relationship is consistent with previously reported observed stability constants (K_{obs}) and was illustrated by Norrild and Eggert in a study of the complexation of *p*-tolylboronic acid with D-glucose. The study provided convincing evidence of a clear preference for the furanose form of the hemiacetal ring at a stoichiometric ratio of 2:1 (boronic acid/D-glucose).[233] The extent of the thermodynamic stability of this complex becomes apparent if one recalls that α-D-glucofuranose accounts for a mere 0.14% of the total speciation of equilibrated D-glucose in water yet becomes the dominant species in solution when complexed to monoboronic acids.[43]

4.10.2 The Preference of Monoboronic Acids for D-Fructose

Drawing these observations together we indicate that the apparent dependence of the stability constants (K) of monoboronic acids on the ability of the vicinal diols of saccharides to conform to a *syn*-periplanar arrangement could be used to explain the heightened stability of D-fructose over other monosaccharides, see Lorand and Edwards' results in Table 1.

The furanose form of D-fructose with an available *syn*-periplanar anomeric hydroxyl pair (C2–C3) is β-D-fructofuranose, Figure 26 (**D**). At equilibrium, this species accounts for an enormous 25% of the total speciation of D-fructose in deuterated water at 31°C.[235] This value can be contrasted with that of the furanose form of D-glucose with an available *syn*-periplanar anomeric hydroxyl pair, α-D-glucofuranose, which accounts for 0.14% of the equilibrated forms of D-glucose in deuterated water at 27°C.[43] In the presence of phenylboronic acid, the stability constants (K) reported by Lorand and Edwards were 4400 M^{-1} with D-fructose and 110 M^{-1} with D-glucose.[64]

From a simplistic viewpoint, we indicated that a very general statistical trend appears to exist between the natural speciation of saccharide forms containing the *syn*-periplanar anomeric hydroxyl pair arrangement (a premise of the form's stability) and the qualitative trend observed for their stability constants (K) with monoboronic acids, Table 2.

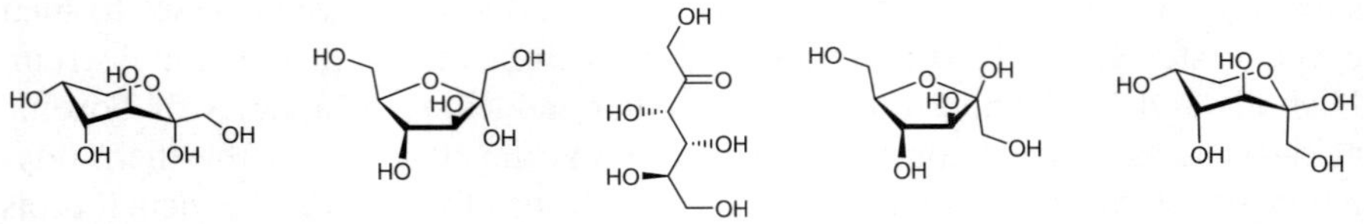

A   **B**   **C**   **D**   **E**

Figure 26   *D-fructose* **39** *in its various configurations and the percentage composition at equilibrium of each form of the sugar in D₂O at 31°C (**A**) α-pyranose, 2.5%; (**B**) α-furanose, 6.5%; (**C**) acyclic form, 0.8%; (**D**) β-furanose, 25%; (**E**) β-pyranose, 65%.[235]*

Table 2   *Saccharide structures containing a syn-periplanar anomeric hydroxyl pair, equilibrated percentages in D₂O and stability constants with phenylboronic acid are tabulated for a selection of saccharides*

Saccharide	Structure	Relative percentage (of the total composition in D₂O)/%	K_{obs} (with phenylboronic acid)[64]/dm³ mol^{-1}
D-glucose (α-D-glucofuranose)	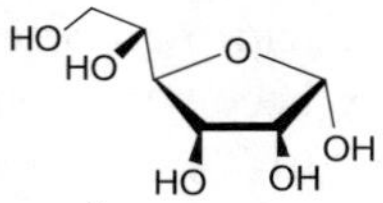	0.14[a]	110
D-mannose (β-D-mannofuranose)	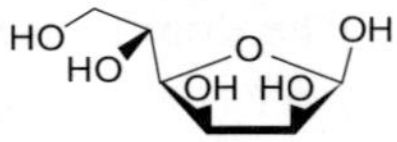	0.3[b]	170
D-galactose (α-D-galactofuranose)	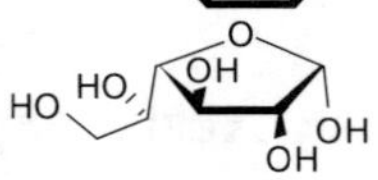	2.5[b]	280
D-arabinose (β-D-arabinofuranose)	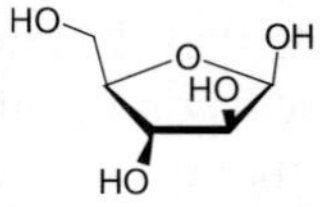	~2[b]	340
D-fructose (β-D-fructofuranose)	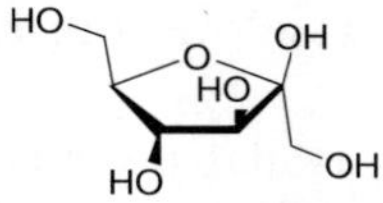	25[b]	4400

[a] Following equilibration in D₂O at 27°C.[43]
[b] Following equilibration in D₂O at 31°C.[235]

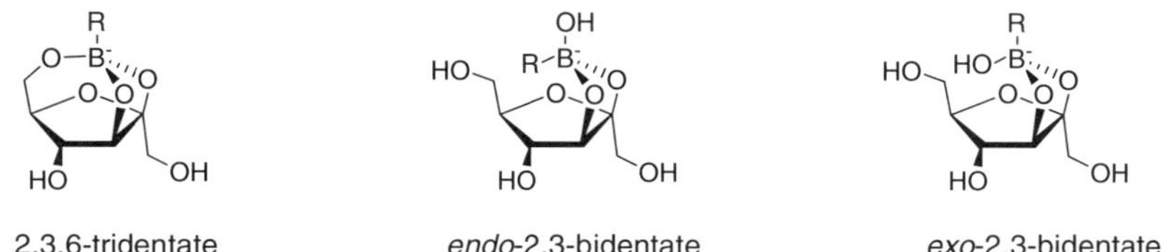

Scheme 33 *The probable twist isomers of the β-D-mannopyranose boronate esters, where R=m-nitro phenylboronic acid.*[234]

Figure 27 *The structures assigned to D-fructose at a 1:1 (p-tolylboronic acid/D-fructose) ratio.*[236]

Unfortunately, the complexity and extensive number of possible binding motifs precludes such a simplistic approach from providing anything other than a statistical rule-of-thumb guide.

The furanose form of a saccharide may not necessarily always be the one favoured for binding. For example, NMR data suggests that D-mannose may be preferentially complexed by monoboronic acids in a pyranose twist conformation, Scheme 33.[234]

Furthermore, bidentate and tridentate complexation is known to occur, with *endo-* and *exo-*orientations of the substituents affecting the thermodynamic stability of the bidentate species, Figure 27.

This was illustrated in an examination into the complexation of *p*-tolylboronic acid with D-fructose.[236] From these two compounds, in aqueous alkaline solution, seven different complexes were afforded. Notably, the abundance of the major species within the system fluctuated as a function of solvent, pH and boronic acid concentration.

4.10.3   Disaccharides

The influence of pyranose to furanose interconversion on the observed stability constants (K_{obs}) of boronic acid complexes can be effectively demonstrated in the fluorescence response of boronic acid appended sensors with disaccharides. In examining the stabilities of disaccharides with the 5-indolylboronic acid **38**, Aoyama and co-workers noted that for non-reducing sugars the observed stability constants (K_{obs}) were significantly reduced or zero. This phenomenon was ascribed to the lack of an anomeric hydroxyl, which inhibited the formation of a strong boronic acid complex with the anomeric hydroxyl and its vicinal neighbour.[139]

135

Cooper extended this investigation by evaluating compound **135** against a range of mono- and di-saccharides.[237] For D-glucose and its disaccharide derivatives melibiose and maltose, strong fluorescence responses were only observed with D-glucose and melibiose (saccharides which can interconvert between their pyranose and furanose forms). Conversely, a weak fluorescence response was observed with maltose, which is covalently bound through its 4-position and as such cannot isomerise to adopt a furanose ring structure (but does posses a free C1,C2-diol functional group).

The same trend was observed for D-fructose and its disaccharide derivatives lactulose and leucrose. Strong fluorescence responses were observed with D-fructose and lactulose (saccharides which can interconvert between their pyranose and furanose forms). Conversely, only a weak fluorescence response was observed with leucrose, which is covalently bound at the 5-position (the fourth carbon as numbered from the anomeric centre) and as such cannot isomerise to adopt a furanose ring structure (but still allows approaching boronic acids unfettered access to the C2,C3-diol).

These results indicate that the negligible observed stability constants (K_{obs}) documented for certain disaccharides cannot be considered solely a function of the availability of the anomeric hydroxyl pair. In the case of the reducing sugars, maltose and leucrose it is the inability of the sugars to adopt their furanose forms, and thus a *syn*-periplanar alignment of the anomeric hydroxyl pair, that inhibits complexation (Schemes 34 and 35).

From the pragmatic viewpoint of designing diboronic acid sensors with selectivity for D-glucose, many receptors have relied on an approximate spacing of the two boronic acid units such that the dimensions of the binding pocket mimic that of other established systems.

While the binding sites for the second boronic acid (at the 3,5,6 or 3,4,6 positions of D-glucose) depend on the experimental conditions and the particular boronic acid being investigated,[220,233] there is agreement that the strongest interaction between monoboronic acids and D-glucose occurs with the *syn*-periplanar hydroxyls on the 1 and 2 positions of α-D-glucofuranose.[139,233,234,237] Given these observations and the known importance of pyranose to furanose interconversion, it has been suggested that in the future design of diboronic acid sensors for saccharides, the pyranose form should not be considered in aqueous systems.[220]

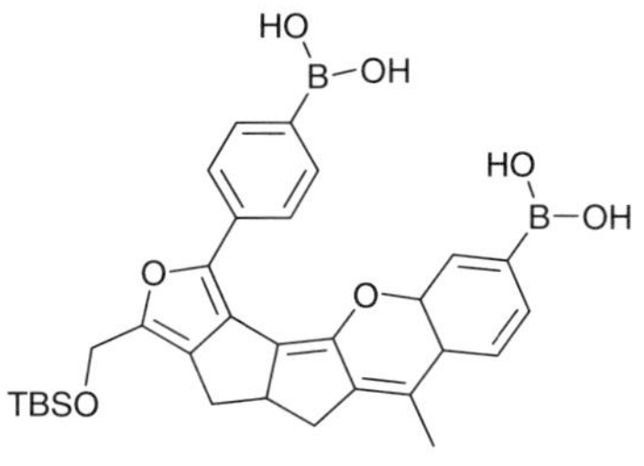

138

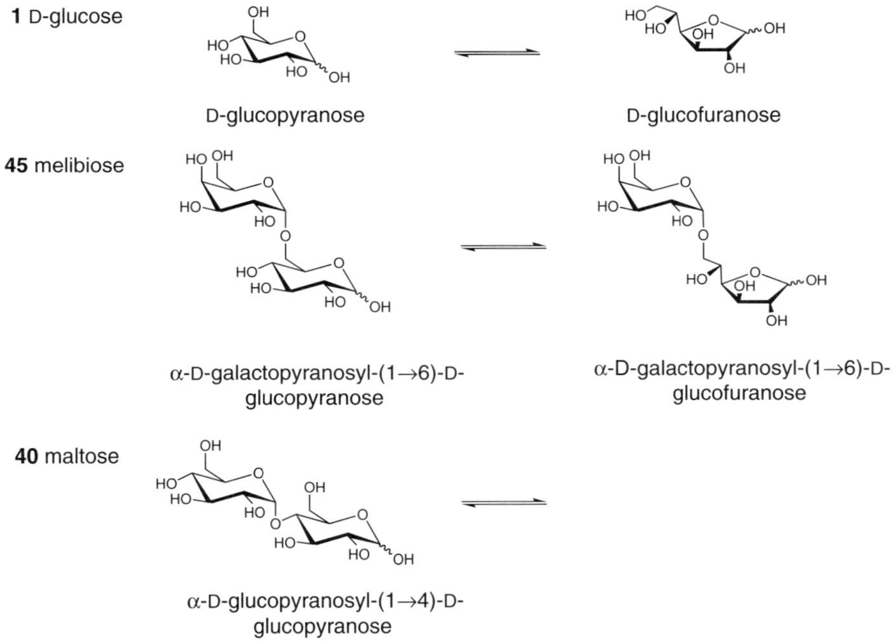

1 D-glucose

D-glucopyranose

D-glucofuranose

45 melibiose

α-D-galactopyranosyl-(1→6)-D-
glucopyranose

α-D-galactopyranosyl-(1→6)-D-
glucofuranose

40 maltose

α-D-glucopyranosyl-(1→4)-D-
glucopyranose

Scheme 34 *D-glucose and its disaccharide derivatives melibiose and maltose.*[237]

Despite these observations Drueckhammer and co-workers successfully used a computer-guided approach to engineer sensor **138**, a receptor specifmshy;ically designed to complex α-D-glucopyranose at the 1,2 and 4,6 positions.[238] The computational approach produced a rigid molecular scaffold anchoring the two boronic acid groups precisely in space. Defined spatial architecture led the receptor to exhibit a 400-fold greater affinity for D-glucose over any of the other saccharides the receptor was evaluated against. Significantly, ^{1}H-NMR confirmed that D-glucose was formed and retained as a stable complex in its pyranose form. Sensor **138** demonstrates that where two-point binding is achieved between boronic acids of fixed distances and enforced geometries, different isomeric forms of a saccharide guest may be witnessed within the binding cleft than would have otherwise been predicted from the simple monoboronic acid structural analogues discussed above.

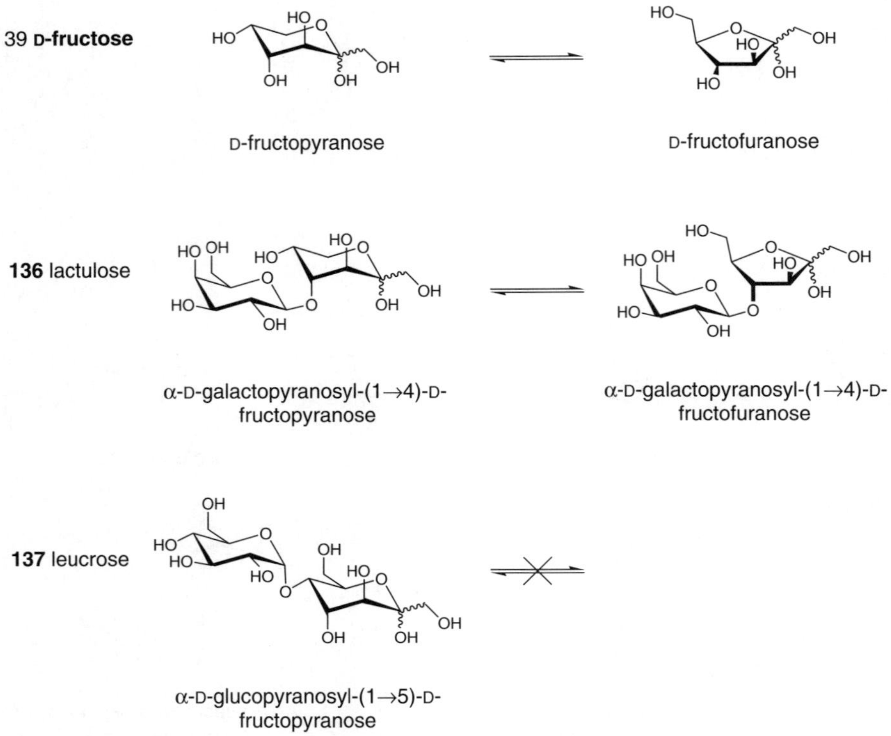

39 D-fructose

D-fructopyranose   D-fructofuranose

136 lactulose

α-D-galactopyranosyl-(1→4)-D-   α-D-galactopyranosyl-(1→4)-D-
fructopyranose   fructofuranose

137 leucrose

α-D-glucopyranosyl-(1→5)-D-
fructopyranose

Scheme 35   *D-fructose and its disaccharide derivatives lactulose and leucrose.*[237]

In refining the selectivity of boronic acid appended sensors for saccharides, it therefore seems that pre-empting the structure of the guest species or the thermodynamic complex that it will form is non-trivial. It is known that only saccharides with the ability to interconvert between pyranose and furanose forms with an available anomeric hydroxyl pair have so far been reported to interact strongly with boronic acids. It is also generally the case that in aqueous solutions the furanose form of the saccharide will be thermodynamically favoured. However, as illustrated by Drueckhammer and others, in the case of recognition sites with two-linked boronic acid fragments, this is a function of substrate structure and geometry.

4.11   Summary

- The primary interaction of a boronic acid with a diol is covalent and involves the rapid and reversible formation of a cyclic boronate ester. Boronic acids are therefore ideally suited to the recognition of saccharides in water, overcoming the problem of solvation inherent in synthetic receptors reliant on hydrogen-bonding interactions.
- On complexation of a vicinal diol to a boronic acid there is a contraction of the O–B–O bond angle with a concomitant increase in the acidity of

the boron species. This phenomenon forms the basis of many of the sensory systems developed around boronic acids.

- Boronic acid–diol complex formation is heavily pH dependent. Rate and stability constants increase by around four and five orders of magnitude, respectively, at pHs above the pK_a of the boronic acid. It has been postulated that these observations can be accounted for by an increase in the rate of the kinetically significant proton transfer mechanism when boron is present as the tetrahedral boronate anion.

- When considering spectrophotometrically derived observed stability constants (K_{obs}) of boronic acids with diols it should be recalled that the observed values have a "medium dependence" on the presence of any Lewis basic components in solution.

- The introduction of a carefully located tertiary amine proximal to the boron centre of a fluorescent sensor permits the sensor to function at lower pH and introduces an "off–on" optical response to the system *via* PET.

- The tertiary amine–boronic acid (N–B) interaction in a boronic acid-based PET sensor has strength in the range of 15–25 kJ mol^{-1}.

- In an aprotic solvent, the N–B dative bond is usually present. However, in a protic media, solvent insertion of the N–B, occurs to afford a hydrogen-bonded zwitterionic species.

- The use of two boronic acid receptors within a binding site introduces saccharide selectivity, permitting saccharides such as D-glucose to be specifically targeted.

- Computational data and observed experimental results indicate that strong binding between boronic acids and saccharides occurs preferentially with saccharides that have an available anomeric hydroxyl pair, which has the capacity to conform to a *syn*-periplanar alignment. In the vast majority of cases, this requires formation of the furanose form of the saccharide.

Modular Fluorescent Sensors

I did not fail 1000 times; the light bulb was developed in 1001 steps
Thomas Alva Edison

During recent years the use of a modular approach in the design and synthesis of new fluorescent PET sensors has been at the fore of the research undertaken towards molecular sensing within our research group.

From a design perspective, fluorescent PET sensors are particularly well suited to a modular construct. The compartmentalisation that is derived from the insulated receptor and fluorophore units means that, where the structure allows, these terminal components can be modified largely independently.

In an effort to understand how recognition trends between boronic acid-based sensors and saccharides are influenced as a function of the sensor's structure, libraries of compounds were synthesised where single variables were modified sequentially. The development of these libraries has been recently reviewed by Phillips and James.[7,‡‡]

5.1 The Design Rationale

In the design of boronic acid based sensory systems it has been established that two receptor units are required if selectivity is to be achieved for specific saccharides.[176] It has been demonstrated that the *N*-methyl-*O*-(aminomethyl)phenylboronic acid fragment is particularly effective at signalling these binding events *via* PET when correctly positioned alongside a suitable fluorophore.[8]

Retaining the same dual boronic acid recognition units throughout, it was thought advantageous to develop a modular system in which the linker and fluorophore units of these sensors could be modified independently. It was postulated that this approach would afford information on the effect of altering the dimensions of the binding pocket, permit the emission wavelength of the systems to be altered and provide clear data on the role of these individual variables by allowing them to be altered in a controlled manner.

In designing a range of modular sensors from which quantitative trends are to be derived, a generic scaffold must be established at the start of the project and adhered to throughout, if a meaningful comparison is to be obtained across the series.

---

‡‡ M. D. Phillips and T. D. James, *J. Fluorescence*, 2004, **14**, 549–559.

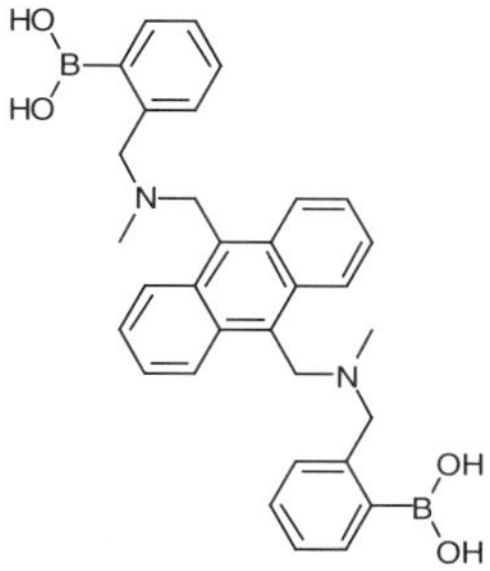

80

While sensors developed around the anthracene core unit (*e.g.* **80**), they have proved to be selective for saccharides such as D-glucose, the rigid core unit (functioning as the scaffold, linker and fluorophore) limits the modifications that can be made to any one part of the system without influencing the sensor as a whole. Perusing the literature it appears that this limitation holds for a great many sensors.

In adopting a modular approach it was deemed important to design a system in which the receptor, linker and fluorophore units could be connected to the molecular scaffold in such a way as to permit these sub-units to be varied independently.

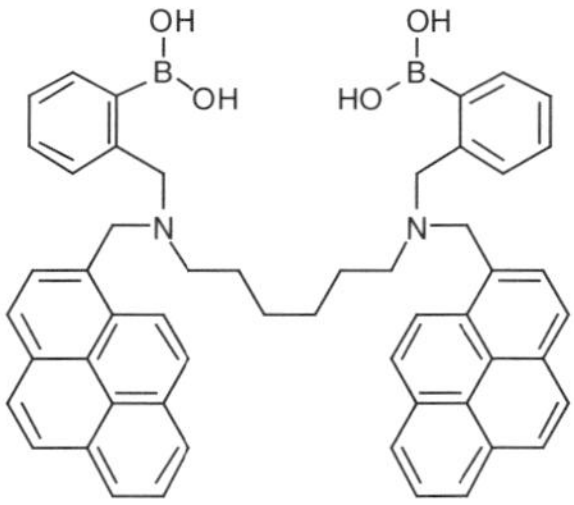

139

Reported in 1995, compound **139** provides an example of a fluorescent PET sensor with clear compartmentalisation of the receptor, linker and fluorophore sub-units.[239] Documented by Sandanayake, James and Shinkai, it was found that the stoichiometry of saccharide binding with sensor **139** (*i.e.* 1:1 or 2:1 saccharide/sensor) could be correlated with a decrease in the fluorescence emission due to the excited state dimer (excimer) complex formed between the two pyrene residues, see Scheme 36 broad peak $\sim$470 nm. Moreover, the saccharide concentration could be monitored *via* the usual increase in fluorescence emission intensity from the LE state of pyrene as a function of PET, see Scheme 36 peaks at 370, 397 and 417 nm. The observed stability constants (K_{obs}) for **139** were 2000 M^{-1} with D-glucose and 790 M^{-1} with D-galactose in 33 wt% methanol in water.[239]

Appleton and Gibson investigated the effect of longer linker units between the two receptor units in this system.[240] 1,6-diaminohexane, 1,7-diaminoheptane, 1,8-diaminooctane, 1,12-diaminododecane and 4,4′-diamino

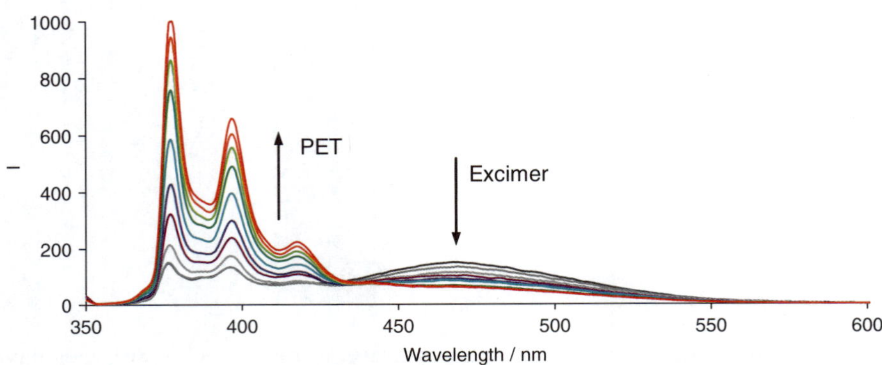

Scheme 36   *The fluorescence emission spectrum of* **139** *(7.5 × 10^{-7} mol dm^{-3}, λ_{ex}=342 nm) with increasing amounts of D-glucose (from 0 to 1.0 × 10^{-1} mol dm^{-3}) in a pH 8.21 aqueous methanolic buffer solution [52.1 wt% methanol (KCl, 0.01000 mol dm^{-3}; KH$_2$PO$_4$, 0.002752 mol dm^{-3}; Na$_2$HPO$_4$, 0.002757 mol dm^{-3})].*

dicyclohexylmethane were introduced, with the observation that six- and seven-carbon linkers conferred the highest selectivity towards D-glucose whereas extended linkers lost the selectivity derived from the cooperative action of the dual recognition units.

Compound **139** was re-synthesised within the James group and the fluorescence properties re-evaluated.[7] The fluorescence titrations of compound **139** (7.5 × 10^{-7} mol dm^{-3}, λ_{ex}=342 nm) with different saccharides were carried out in a pH 8.21 aqueous methanolic buffer solution [52.1 wt% methanol (KCl, 0.01000 mol dm^{-3}; KH$_2$PO$_4$, 0.002752 mol dm^{-3}; Na$_2$HPO$_4$, 0.002757 mol dm^{-3})]. The observed stability constants (K_{obs}) with coefficient of determination (r^2) were calculated by the fitting of emission intensity *vs.* saccharide concentration using custom written non-linear (Levenberg–Marquardt algorithm) curve fitting.[191] The errors reported are the standard errors obtained from the best fit. Relative fluorescent enhancements (I/I$_0$) are also reported. The results for **139** with D-glucose were K_{obs}=260 ± 15 M^{-1} (r^2=1.00), I/I$_0$=4.9; with D-galactose K_{obs}=237 ± 6 M^{-1} (r^2=1.00), I/I$_0$=4.2; with D-fructose K_{obs}=244 ± 26 M^{-1} (r^2=0.99), I/I$_0$=3.4 and with D-mannose K_{obs}=32 ± 3 M^{-1} (r^2=0.99), I/I$_0$=3.2.

In Scheme 36 the long wavelength excimer emission due to π stacking of the pyrene fluorophores can be observed around 470 nm. This fluorescence emission has a somewhat more modest intensity than the emission observed due to the LE state with peaks at 370, 397 and 417 nm. As D-glucose binds to the receptor it forms a rigid 1:1 cyclic structure and in so doing increases the proximity of the boronic acids. It is thought that this interaction levers the two pyrene fluorophores apart so as to increase their separation and reduce the π–π interactions between them. This change in geometry of the system's components manifests itself in the fluorescence emission spectrum as a clear decrease in excimer emission intensity with added D-glucose.

The increase in the emission intensity from the LE state with increased saccharide concentration, modulated by PET, remains largely unaffected when compared to sensors with single fluorophores.

5.2 Modular Systems

Six carbons separate the two amino nitrogens in the D-glucose selective compound **139**, the same separate the two amino nitrogens in the D-glucose selective sensor **80**. The boronic acid groups appended to the two nitrogens of sensor **139** (and **80**) must both take part in a binding event to allow the full fluorescence emission of the system to be restored. The dual recognition sites and their separation are therefore key to the observed saccharide selectivity. With this in mind it was decided that a generic template based on sensor **139** would prove effective. The general design assembly proposed is illustrated in Figure 28.

This design retains the two boronic acid groups required for selectivity but allows the separation between them to be varied by altering the linker. It also permits the fluorophore to be varied independently and by using only one fluorophore overcomes the problems that may arise from excimer emission,[§§] insolubility, excessive hydrophobicity and steric crowding at the binding pocket.

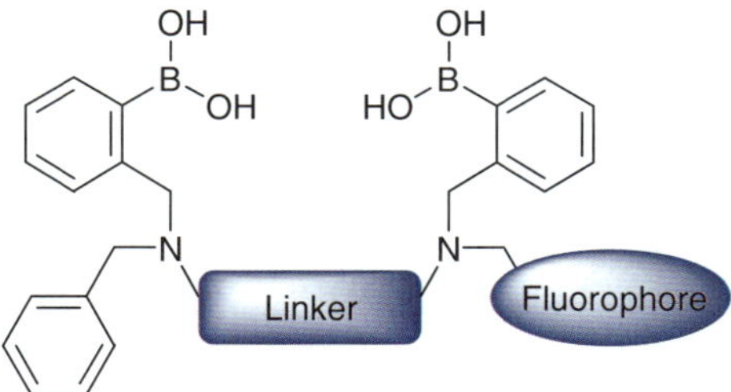

Figure 28 *Receptor and spacer groups remained unchanged throughout. Two N-methyl-O-(aminomethyl)phenylboronic acid groups functioned as the receptor units with a methylene spacer modulating PET.*

5.2.1 Linker Dependence

The modular PET sensors **140**$_{(n=3)}$–**145**$_{(n=8)}$ contained two phenylboronic acid groups, a pyrene fluorophore and a variable linker (Figure 29). The linker was varied from trimethylene **140**$_{(n=3)}$ to octamethylene **145**$_{(n=8)}$. The aim of this research being to elucidate the optimum linker length required for maximum D-glucose selectivity.[241–243]

[§§] While excimer emission can be used to visually determine the stoichiometry of the binding event and advantageously red shifts the fluorescence emission, the comparatively small changes in fluorescence emission intensity do not lend this part of the emission band to accurate signalling of saccharide concentrations. It is also generally the case that sensors with two fluorophore units have observed binding constants (K_{obs}) ~four times lower than their monofluorophore counterparts.

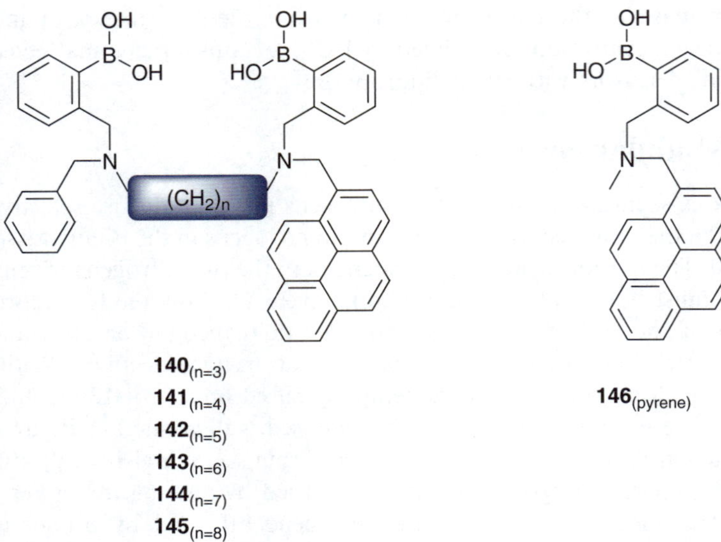

Figure 29 *The modular PET sensors **140**(n=3)–**145**(n=8) with monoboronic acid reference compound **146**(pyrene).*

The fluorescence titrations of **140**$_{(n=3)}$–**145**$_{(n=8)}$ and **146**$_{(pyrene)}$ (1.0×10^{-7} mol dm^{-3}) with different saccharides were carried out in a pH 8.21 aqueous methanolic buffer solution, as described above (see page 86).[244] The fluorescence intensity of sensors **140**$_{(n=3)}$–**145**$_{(n=8)}$ and **146**$_{(pyrene)}$ increased with increasing saccharide concentration. The observed stability constants (K_{obs}) of PET sensors **140**$_{(n=3)}$–**145**$_{(n=8)}$ and **146**$_{(pyrene)}$ were calculated by fitting the emission intensity at 397 nm *vs.* saccharide concentration and are given in Table 3.

To help visualise the trends in the observed stability constants (K_{obs}) documented in Table 3, the observed stability constants (K_{obs}) of the diboronic acid sensors **140**$_{(n=3)}$–**145**$_{(n=8)}$ are reported in Figure 30 divided by (*i.e.* relative to) the observed stability constants (K_{obs}) of their equivalent monoboronic acid analogue **146**$_{(pyrene)}$. In most cases, the observed stability constants (K_{obs}) with diboronic acid sensors **140**$_{(n=3)}$–**145**$_{(n=8)}$ are higher than for the monoboronic acid sensor **146**$_{(pyrene)}$.

It is known that D-glucose and D-galactose will bind to diboronic acids readily using two sets of diols, thus forming stable, cyclic 1:1 complexes. The allosteric binding of the two boronic acid groups is clearly illustrated by the relative difference between the observed stability constants (K_{obs}) of the equivalent di- and monoboronic acid compounds. The observed stability constants (K_{obs}) of the diboronic acid sensors are up to $\sim$22 times larger than with their monoboronic acid counterparts, see Figure 30.

The observed stability constants (K_{obs}) for the diboronic acid sensors with D-fructose and D-mannose are, at most, twice as strong as with the monoboronic acid sensor **146**$_{(pyrene)}$. Each D-fructose and D-mannose molecule will only bind to one boronic acid unit through one diol. This allows complexes to form with

Table 3 *Quantum yield (qFM) values for compounds $140_{(n=3)}$–$145_{(n=8)}$ and $146_{(pyrene)}$ in the absence of saccharides. Observed stability constant (K_{obs}) (determination of coefficient, r^2) and fluorescence enhancements for compounds $140_{(n=3)}$–$145_{(n=8)}$ and $146_{(pyrene)}$ with monosaccharides. The system with the highest observed stability constant (K_{obs}) is highlighted in red*

Sensor	qFM	D-Glucose		D-Galactose	
		$K_{obs}/dm^3\ mol^{-1}$	Fluorescence enhancement	$K_{obs}/dm^3\ mol^{-1}$	Fluorescence enhancement
$140_{(n=3)}$	0.16	103 ± 3 (1.00)	3.9	119 ± 5 (1.00)	3.5
$141_{(n=4)}$	0.16	295 ± 11 (1.00)	3.3	222 ± 17 (1.00)	3.7
$142_{(n=5)}$	0.20	333 ± 27 (1.00)	3.4	177 ± 15 (1.00)	3.0
$143_{(n=6)}$	0.24	962 ± 70 (0.99)	2.8	657 ± 39 (1.00)	3.1
$144_{(n=7)}$	0.16	336 ± 30 (0.98)	3.0	542 ± 41 (0.99)	2.9
$145_{(n=8)}$	0.19	368 ± 21 (1.00)	2.3	562 ± 56 (0.99)	2.3
$146_{(pyrene)}$	0.17	44 ± 3 (1.00)	4.5	51 ± 2 (1.00)	4.2

Sensor	qFM	D-Fructose		D-Mannose	
		$K_{obs}/dm^3\ mol^{-1}$	Fluorescence enhancement	$K_{obs}/dm^3\ mol^{-1}$	Fluorescence enhancement
$140_{(n=3)}$	0.16	95 ± 9 (0.99)	3.6	45 ± 4 (1.00)	2.7
$141_{(n=4)}$	0.16	266 ± 28 (0.99)	4.2	39 ± 1 (1.00)	3.4
$142_{(n=5)}$	0.20	433 ± 19 (1.00)	3.4	48 ± 2 (1.00)	3.0
$143_{(n=6)}$	0.24	784 ± 44 (1.00)	3.2	74 ± 3 (1.00)	2.8
$144_{(n=7)}$	0.16	722 ± 37 (1.00)	3.3	70 ± 5 (1.00)	2.7
$145_{(n=8)}$	0.19	594 ± 56 (0.99)	2.3	82 ± 3 (1.00)	2.2
$146_{(pyrene)}$	0.17	395 ± 11 (1.00)	3.6	36 ± 1 (1.00)	3.7

overall 2:1 (saccharide/sensor) stoichiometry. The relative values of *ca.* 2 are indicative of two independent saccharide binding events on each sensor, with no increase in stability derived from co-operative binding.

The highest observed stability constants (K_{obs}) for D-glucose within these systems were obtained by sensor $143_{(n=6)}$. The flexible six carbon linker provided the optimal selectivity for D-glucose over other saccharides, this is in agreement with the observed selectivity of compounds **80** and **139**, which also have linkers containing six carbon atoms.

Curiously, there is an inversion in the selectivity displayed by these systems on going from a six to a seven carbon linker. From Figure 30 it is clear that the trimethylene linked $140_{(n=3)}$ shows little specificity between D-glucose and D-galactose. Increasing the size of the binding pocket, tetramethylene $141_{(n=4)}$ through to hexamethylene $143_{(n=6)}$, induces a clear selectivity for D-glucose, with $143_{(n=6)}$ providing the strongest binding by far. However, there is an inversion in this selectivity on increasing the linker length to heptamethylene

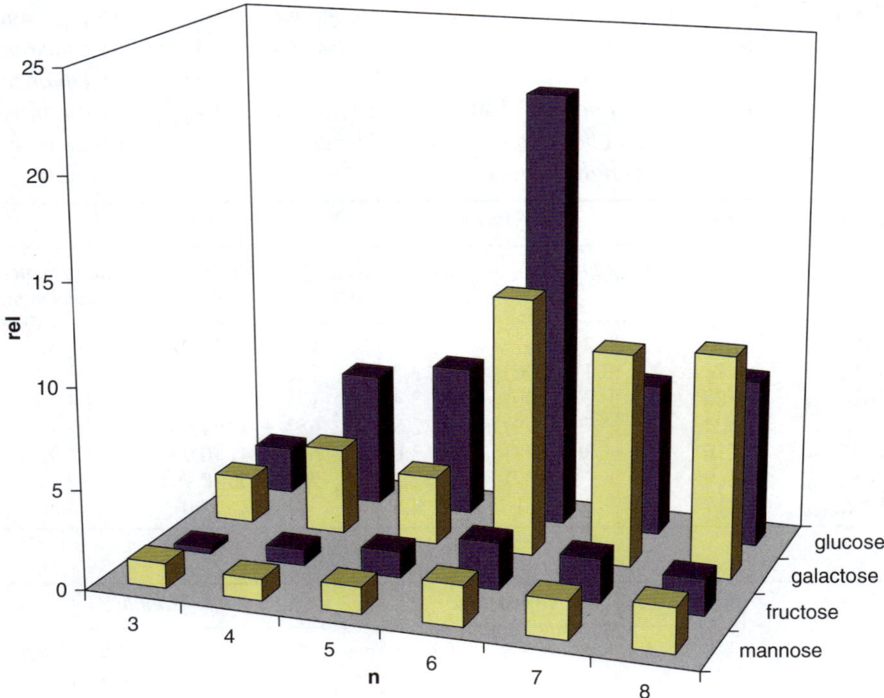

Figure 30   *Observed stability constants (K_{obs}) of **140**(n=3)–**145**(n=8) divided by the observed stability constants (K_{obs}) of **146**(pyrene), to yield relative values with saccharides. The D-configuration of the monosaccharides was used throughout this evaluation.*

144$_{(n=7)}$ and octamethylene **145**$_{(n=8)}$, with the enlarged binding pocket being D-galactose selective.

This facet was initially ascribed to the difference in configuration of these diastereomers.[243] The 1,2- and 4,6-diols of D-glucose point in the same direction (down), but in D-galactose (the 4-epimer of D-glucose), the 1,2-diol is down and the 4,6-diol is up, this is illustrated in Figure 31. Therefore, not only are the interdiol distances of D-galactose slightly longer than in D-glucose but the diols are also on opposite faces of the saccharide ring. On this basis it seemed reasonable to expect better complementarity for D-galactose with a larger inter-receptor distance than for D-glucose.

As discussed above (see Section 4.10, *The Importance of Pyranose to Furanose Interconversion*, page 75) current thinking requires that we consider saccharidic forms where a *syn*-periplanar arrangement of the anomeric hydroxyl pair can be attained. Generally, this requires formation of the furanose form of the saccharide. However, computational work has shown that in the case of D-galactose the α-D-furanose form of the saccharide is not the only species that can be considered with a syn-periplanar alignment of the anomeric hydroxyl pair[234] (Figure 32).

Figure 31   *The α-D-pyranose forms of glucose and galactose with the 1,2- and 4,6-diols highlighted in red.*

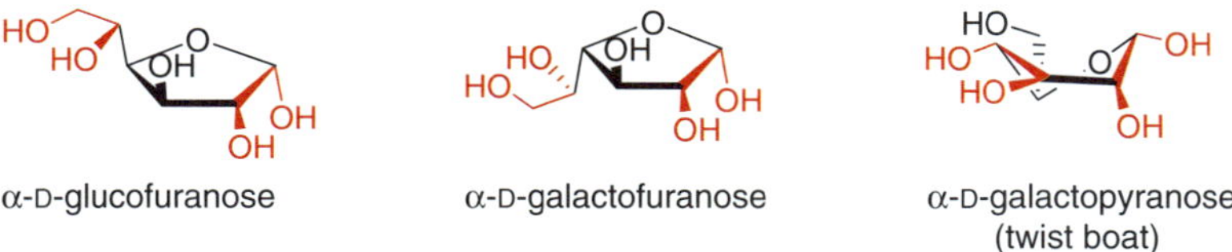

Figure 32   *The α-D-furanose forms of glucose and galactose with the 1,2- and 5,6-diols highlighted in red, as well as the twist boat conformation of the α-D-pyranose form of galactose with the 1,2- and 3,4-diols highlighted in red.*

On complexation of a boronic acid to the diol in the 1,2-position of α-D-galactopyranose a re-orientation of the saccharide can afford an acetal ring with a boat or twist boat conformation. This re-orientation permits a second boronic acid group to bind with the diol in the 3,4-position producing an energetically stable complex with the pyranose form of the saccharide.

In light of this new data it seems plausible that the preference of D-galactose for a slightly larger binding pocket than D-glucose can be ascribed to the stability of the preferred ring form. D-galactose can generate complexes with boronic acids in its pyranose form, whereas D-glucose will prefer the smaller furanose ring form.

5.2.2   Linker Dependence and Disaccharides

When compared to monosaccharide receptors only a small number of synthetic, boronic acid based receptors for di-[53,214,219,245] and oligosaccharides[246–253] currently exist. One reason is the problem associated with positioning the two binding units to create a selective receptor for these larger and more flexible guests. Following the investigation into the effect of linker length on monosaccharides, the selectivity of sensors $140_{(n=3)}$–$145_{(n=8)}$ towards disaccharides was probed, see Arimori, Phillips and James.[254,¶¶]

For consistency the same disaccharides were evaluated here as were used in the evaluation of sensor **135**.[237] For the structures of D-glucose, melibiose and maltose, see Scheme 34 and for the structures of D-fructose, lactulose and leucrose, see Scheme 35. The fluorescence titrations of sensors $140_{(n=3)}$–$145_{(n=8)}$ and $146_{(pyrene)}$ with different saccharides, were carried out in a pH 8.21 aqueous methanolic buffer solution, as described earlier. The fluorescence intensity of $140_{(n=3)}$–$145_{(n=8)}$ and $146_{(pyrene)}$ (1.0×10^{-7} mol dm^{-3}, λ_{ex}=342 nm) increased

---

¶¶S. Arimori, M.D. Phillips and T.D. James, *Tetrahedron Lett.*, 2004, **45**, 1539–1542.

with increasing saccharide concentration. The observed stability constants (K_{obs}) of PET sensors $140_{(n=3)}$–$145_{(n=8)}$ and $146_{(pyrene)}$ were calculated by fitting the emission intensity at 397–nm *vs.* saccharide concentration curves. Fluorescence enhancements were calculated by the same method. Where the emission intensity at 397 nm *vs.* concentration of saccharide curves failed to describe a distinct plateau the maximum observed fluorescence enhancements were reported. The observed stability constants (K_{obs}) and fluorescence enhancements for $140_{(n=3)}$ $145_{(n=8)}$ and $146_{(pyrene)}$ are given in Table 4 and are displayed graphically in Figures 33 and 34.

The observed stability constants (K_{obs}) for compounds $140_{(n=3)}$–$145_{(n=8)}$ and $146_{(pyrene)}$ with saccharides, which have the ability to form a furanose ring with a *syn*-periplanar anomeric hydroxyl pair (D-glucose, melibiose, D-fructose and lactulose) are large, consistent with our understanding of the boronic acid diol interaction. Equally, the low or zero values of the observed stability constants (K_{obs}) for saccharides unable to form a furanose ring with a *syn*-periplanar anomeric hydroxyl pair (maltose and leucrose) are also consistent with current thinking.

Table 4 *Observed stability constant (K_{obs}) (determination of coefficient, r^2) and fluorescence enhancement for the disaccharide complexes of compounds $140_{(n=3)}$–$145_{(n=8)}$ and $146_{(pyrene)}$[a]*

Sensor	Melibiose		Maltose	
	$K_{obs}/dm^3\,mol^{-1}$	Fluorescence enhancement	$K_{obs}/dm^3\,mol^{-1}$	Fluorescence enhancement
$140_{(n=3)}$	33 ± 7 (0.98)	2.4	0 ± 2 (0.96)	1.5[b]
$141_{(n=4)}$	77 ± 9 (0.99)	4.9[b]	31 ± 7 (0.98)	3.5[b]
$142_{(n=5)}$	184 ± 11 (1.00)	3.8	2 ± 1 (0.99)	2.1[b]
$143_{(n=6)}$	339 ± 17 (1.00)	2.9	52 ± 14 (0.98)	2.3
$144_{(n=7)}$	153 ± 4 (1.00)	3.3	5 ± 1 (1.00)	2.0[b]
$145_{(n=8)}$	192 ± 30 (0.98)	2.9	22 ± 4 (0.98)	2.4[b]
$146_{(pyrene)}$	96 ± 5 (1.00)	3.9	5 ± 1 (0.99)	2.1[b]

Sensor	Lactulose		Leucrose	
	$K_{obs}/dm^3\,mol^{-1}$	Fluorescence enhancement	$K_{obs}/dm^3\,mol^{-1}$	Fluorescence enhancement
$140_{(n=3)}$	126 ± 14 (0.99)	3.5	29 ± 3 (0.99)	2.5[b]
$141_{(n=4)}$	477 ± 92 (0.95)	4.5	35 ± 2 (1.00)	3.4[b]
$142_{(n=5)}$	616 ± 114 (0.97)	5.0	41 ± 2 (1.00)	3.5[b]
$143_{(n=6)}$	595 ± 30 (1.00)	3.2	69 ± 2 (1.00)	2.9[b]
$144_{(n=7)}$	493 ± 28 (1.00)	3.3	21 ± 5 (0.97)	3.9[b]
$145_{(n=8)}$	528 ± 29 (1.00)	2.6	72 ± 5 (1.00)	2.7[b]
$146_{(pyrene)}$	473 ± 10 (1.00)	3.7	58 ± 3 (1.00)	3.7

[a] The K values were analysed in KaleidaGraph using nonlinear (Levenberg–Marquardt algorithm) curve fitting. The errors reported are the standard errors obtained from the best fit.
[b] Maximum observed fluorescence enhancement.

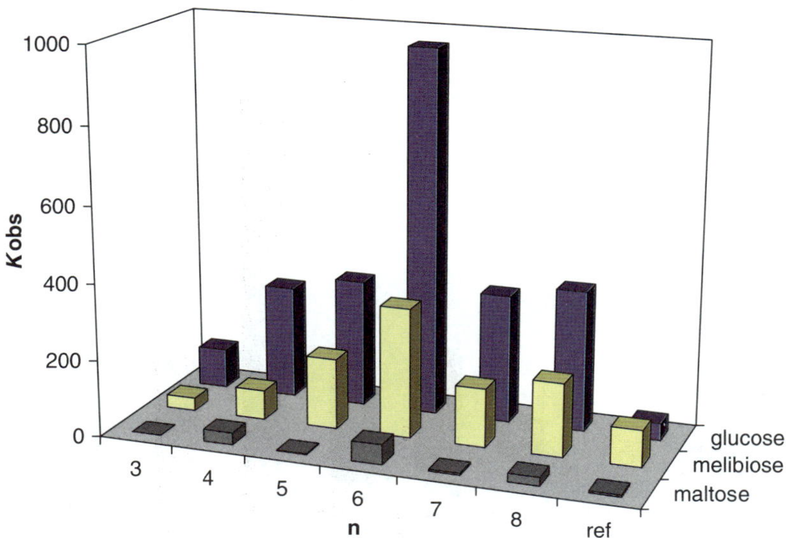

Figure 33 *Observed stability constants (K_{obs}) of the diboronic acid sensors **140**(n=3)–**145**(n=8) and the monoboronic acid reference compound **146**(pyrene) (denoted n=ref) with the D-configuration of glucose and its derivatives.*

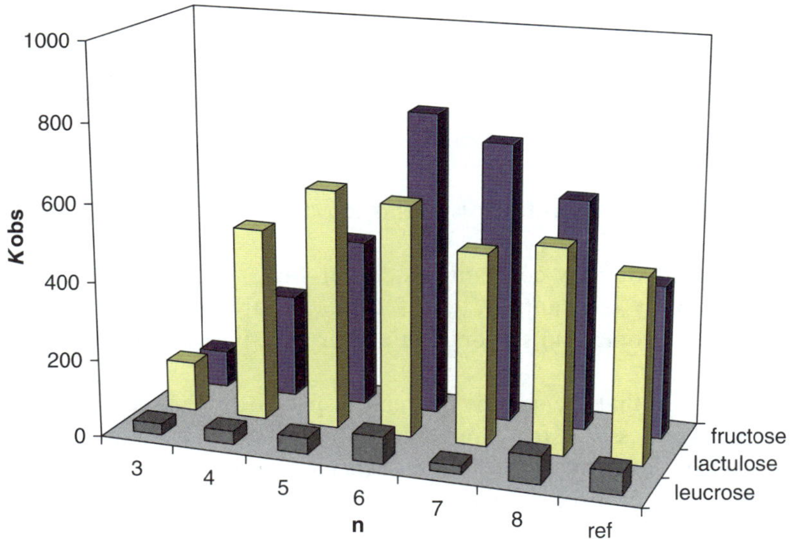

Figure 34 *Observed stability constants (K_{obs}) of the diboronic acid sensors **140**(n=3)–**145**(n=8) and the monoboronic acid reference compound **146**(pyrene) (denoted n=ref) with the D-configuration of fructose and its derivatives.*

The importance of this work compared to previous evaluations is that it was the first to probe the efficacy of diboronic acid units in disaccharide recognition. From previous results it was apparent that the second binding site could bind monosaccharides at their 3, 4, 5 or 6 positions. What had not

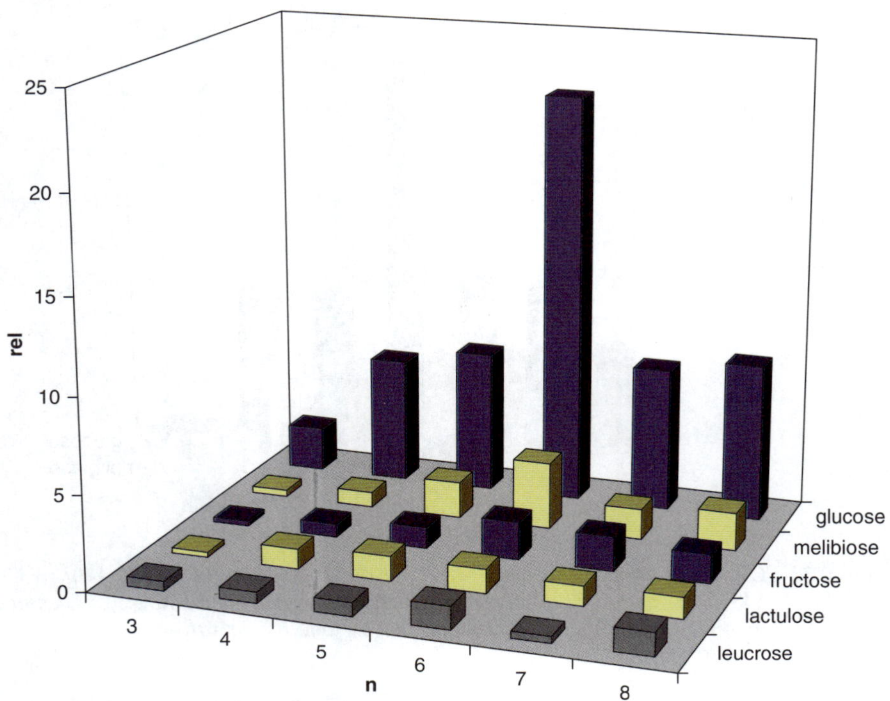

Figure 35   *Observed stability constants (K_{obs}) of **140**(n=3)–**145**(n=8) divided by the observed stability constants (K_{obs}) of **146**(pyrene), to yield relative values with saccharides. The D-configuration of the monosaccharides was used throughout this evaluation.*

been examined before was the role of this secondary binding site with disaccharides.

The observed stability constants (K_{obs}) for **140**$_{(n=3)}$–**145**$_{(n=8)}$ divided by the observed stability constants (K_{obs}) of **146**$_{(pyrene)}$ are displayed in Figure 35. Overall the diboronic acid sensors retain their selectivity for D-glucose over the other saccharides.

It can be seen that the selectivity trend displayed by the sensors towards D-glucose is, to a lesser extent, mirrored by melibiose. An examination of the furanose ring segment of melibiose, Scheme 37, reveals that binding was feasible at the 3,5-positions allowing for the potential formation of a stable cyclic complex.

The relative values obtained for maltose were not illustrated in Figure 35 as the relative values were found to fluctuate significantly. The cause of this lies in the value obtained for the observed stability constant (K_{obs}) of reference compound **146**$_{(pyrene)}$. As the observed stability constant (K_{obs}) of compound **146**$_{(pyrene)}$ with maltose was only 5 $\pm$ 1 M^{-1} when the relative values were calculated small changes in the observed stability constants (K_{obs}) were found to manifest large changes in the relative values. Nonetheless, it should be noted that observed stability constants (K_{obs}) for sensors **140**$_{(n=3)}$–**145**$_{(n=8)}$ with

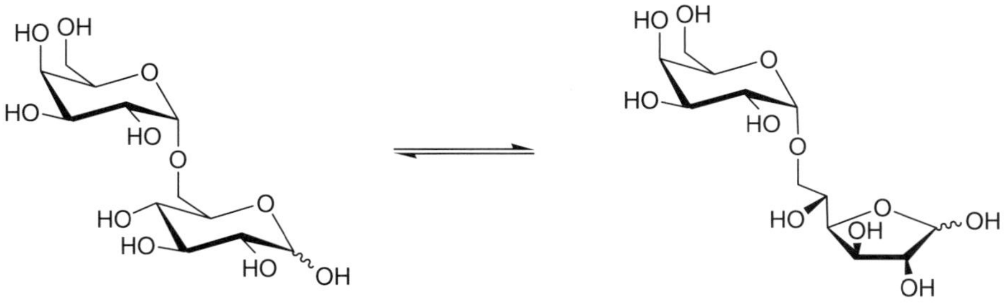

Scheme 37   *The pyranose to furanose interconversion of the glucose ring in melibiose.*

maltose were small in all cases, with all values contained within the range 0 ± 2 M^{-1} to 52 ± 14 M^{-1}.

It is clear from the data that where the secondary binding sites are blocked on saccharides *i.e.* on maltose, lactulose and leucrose, no increased stability is derived from the introduction of a second boronic acid unit.

Overall selectivity is retained for the monosaccharides with the ability to form cyclic complexes. An analogous trend is observed for disaccharides, which also have the ability to form cyclic complexes, however, the observed stability constants (K_{obs}) are weaker in these instances, a characteristic attributed to the increased steric hindrance at the secondary binding site.

5.3   Energy Transfer Systems

Fluorescence energy transfer is the transfer of excited state energy from a donor to an acceptor. The transfer occurs as a result of transition dipole–dipole interactions between the donor–acceptor pair.[128] The fluorescent sensor investigated, **147**(phenanthrene–pyrene), had two phenylboronic acid groups a hexamethylene linker and two different fluorophore groups: phenanthrene and pyrene.[255]

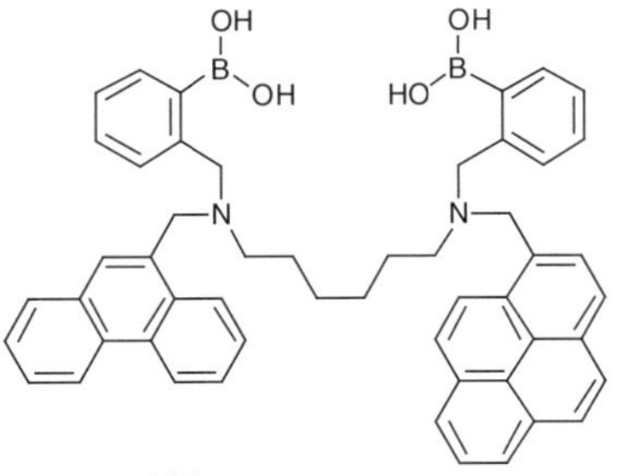

147(phenanthrene-pyrene)

The purpose of constructing sensor **147**(phenanthrene–pyrene) was to investigate the efficiency of energy transfer from phenanthrene to pyrene as a function of saccharide binding. A similar concept had previously been employed in the construction of a fluorescent calix[4]arene sodium sensor.[256] The excitation and emission wavelengths of the phenanthrene (donor) are 299 and 369 nm, respectively, while the excitation and emission wavelengths of the pyrene (acceptor) are 342 and 397(or 417) nm, respectively. The emission wavelength of phenanthrene (369 nm) and excitation wavelength of pyrene (342 nm) overlap. These observations led to the postulation that intramolecular energy transfer from phenanthrene to pyrene could take place in sensor **147**$_{(phenanthrene–pyrene)}$.

Using light of 342 nm (the λ_{ex} of pyrene) to excite sensor **147**$_{(phenanthrene–pyrene)}$ (Scheme 38) leads to long wavelength excimer emission due to π–π interactions between phenanthrene and pyrene, Scheme 39 broad peak $\sim$470 nm. PET modulated emission is also observed from the LE state of pyrene, the

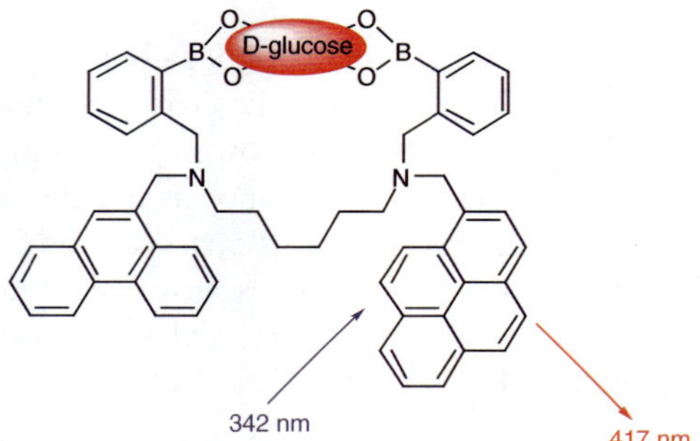

Scheme 38 *Schematic representation of sensor* **147**$_{(phenanthrene–pyrene)}$ *with bound D-glucose (depicted as a red ellipse). D-Glucose allows a revival of fluorescence emission from the LE pyrene fluorophore when pyrene is excited directly. Excitation and emission wavelengths are illustrated in blue and red, respectively. The colours are used to depict the Stokes shift of the system rather than the actual colour of the light.*

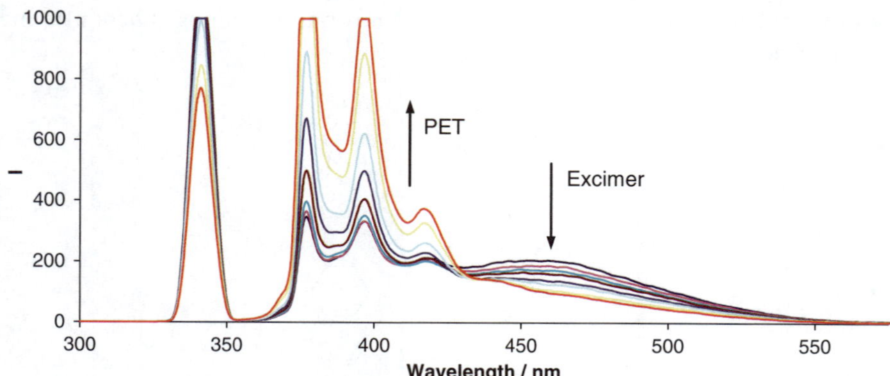

Scheme 39 *Fluorescence intensity vs. wavelength for sensor* **147**$_{(phenanthrene–pyrene)}$ *(2.5 × 10^{-6} mol dm^{-3}) displaying photoinduced electron transfer and excimer emission with increasing concentrations of D-glucose (from 0 to 1.0 × 10^{-1} mol dm^{-3}) in 52.1 wt% MeOH pH 8.21 phosphate buffer. λ_{ex}=342 nm.*

three distinctive peaks of the emission band visible in Scheme 39 at 370, 397 and 417 nm. The intense peak in Scheme 39 at 342 nm is due to Raleigh scattering of the exciting radiation.

If light of 299 nm was used (the λ_{ex} of phenanthrene) to excite sensor **147**(phenanthrene–pyrene) the fluorescence emission spectrum did not display an emission band around 369 nm as would be expected for phenanthrene. Instead the resulting emission band had an emission maximum at 417 nm and was identical to the emission observed when pyrene was excited directly, as in Scheme 39. This result can be ascribed to energy transfer from the phenanthrene to the pyrene fluorophore (Schemes 40 and 41).

The fluorescence titrations of sensor **147**(phenanthrene–pyrene) (2.5×10^{-6} mol dm^{-3}, λ_{ex}=299 nm for phenanthrene and λ_{ex}=342 nm for pyrene) with different saccharides were carried out in a pH 8.21 aqueous methanolic buffer, as described above (page 86). Absorption *vs.* concentration plots of sensor **147**(phenanthrene–pyrene) and the monoboronic acid reference compounds **146**(pyrene) and **153**(phenanthrene) confirmed that the π–π stacking of sensor **147**(phenanthrene–pyrene) was solely intramolecular. The fluorescence intensity of sensor **147**(phenanthrene–pyrene) at 417 nm increased with added saccharide when excited at both 299 and 342 nm, while the excimer emission at 460 nm decreased with added saccharide. The change in excimer emission indicates that the π–π interaction between phenanthrene and pyrene is disrupted on saccharide binding.

The observed stability constants (K_{obs}) of sensor **147**(phenanthrene–pyrene) (with λ_{ex}=299 nm and λ_{ex}=342 nm) were calculated by fitting the emission intensities at 417 nm *vs.* concentration of saccharide curves and are given in Table 5. The observed stability constants (K_{obs}) for the diboronic acid sensor **147**(phenanthrene–pyrene) (λ_{ex}=299 and 342 nm) with D-glucose were enhanced relative to those of the monoboronic acid reference compounds **146**(pyrene) and **153**(phenanthrene), while the observed stability constants (K_{obs}) for the diboronic acid sensor

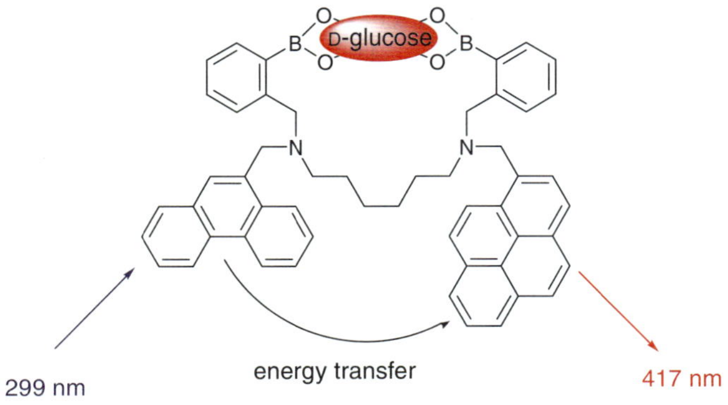

Scheme 40 *Schematic representation of sensor **147**(phenanthrene–pyrene) with bound D-glucose (depicted as a red ellipse). When phenanthrene is excited directly an energy transfer mechanism can occur. Excitation and emission wavelengths are illustrated in blue and red, respectively. The colours are used to depict the Stokes shift of the system rather than the actual colour of the light. The non-radiative energy transfer is depicted in black.*

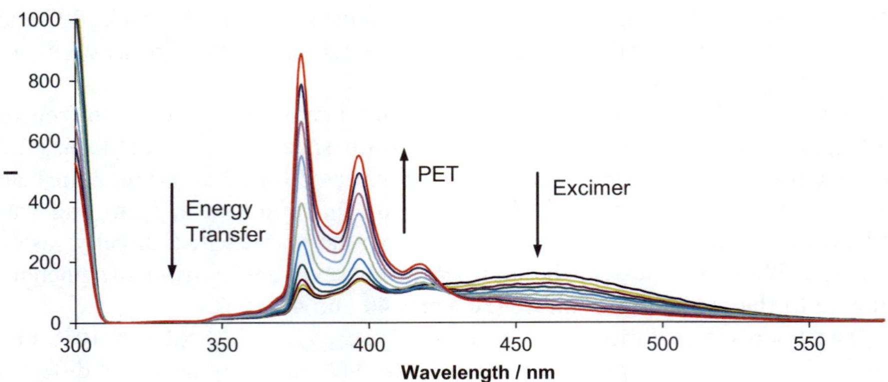

Scheme 41   *Fluorescence intensity vs. wavelength for sensor* **147**$_{(phenanthrene–pyrene)}$ *(2.5 ×*
10^{-6} mol dm^{-3}) displaying energy transfer, photoinduced electron transfer and
excimer emission with increasing concentrations of D-*glucose (from 0 to 1.0 ×*
10^{-1} mol dm^{-3}) in 52.1 wt% MeOH pH 8.21 phosphate buffer. λ$_{ex}$=299 nm.

147$_{(phenanthrene–pyrene)}$ (λ$_{ex}$=299 and 342 nm) with D-fructose were reduced
relative to those for the monoboronic acid reference compounds **146**$_{(pyrene)}$
and **153**$_{(phenanthrene)}$.

Sensor **147**$_{(phenanthrene–pyrene)}$ was particularly noteworthy in that the difference
between the observed fluorescence enhancements obtained when excited at
phenanthrene (λ$_{ex}$=299 nm) and pyrene (λ$_{ex}$=342 nm) can be correlated with
the molecular structure of the saccharide-sensor complex *via* PET, see Table 5.
The fluorescence enhancement of sensor **147**$_{(phenanthrene–pyrene)}$ with D-glucose was
3.9 times greater when excited at 299 nm and 2.4 times greater when excited at
342 nm. With D-fructose the enhancement was 1.9 times greater when
excited at 299 nm and 3.2 times greater when excited at 342 nm. This is to say
that the selectivity between D-fructose and D-glucose within the same complex is
in fact inverted dependant on the excitation wavelength used. This inversion is
clearly displayed in Figures 36 and 37. This result arises from the fact that energy
transfer from phenanthrene to pyrene is far more efficient in the 1:1 cyclic
diboronic acid – saccharide complex than in the alternative 1:2 acyclic complex.
This approach finally allows for efficient discrimination between saccharides
based on their binding motif (*i.e.* the formation of a 1:1 or a 1:2 complex).

5.4   Fluorophore Dependence in Modular Systems

Given the understanding developed in the previous chapter regarding the
influence of the linker length in modular PET sensors for saccharides, as well
as the advantages of using energy transfer to determine selectivity, the next
structural feature to be evaluated was the fluorophore appended to the modular
sensor, see Arimori, Consiglio, Phillips and James.[257,‖‖]

‖‖ S. Arimori, G. A. Consiglio, M. D. Phillips, and T. D. James, *Tetrahedron Lett.*, 2003, **44**,
4789–4792.

Table 5 *Observed stability constants (K_{obs}) (coefficient of determination, r^2) and fluorescence enhancements for compounds* **139** *and* **147**$_{(phenanthrene-pyrene)}$ *with saccharides*[a]

Sensor	D-Glucose		D-Galactose	
	$K_{obs}/dm^3\ mol^{-1}$	Fluorescence enhancement	$K_{obs}/dm^3\ mol^{-1}$	Fluorescence enhancement
139	260 ± 15 (1.00)	4.9	237 ± 6 (1.00)	4.2
147[b]	142 ± 12 (0.99)	3.9	74 ± 7 (0.99)	2.2
147[d]	108 ± 10 (0.99)	2.4	81 ± 8 (0.99)	2.6

Sensor	D-Fructose		D-Mannose	
	$K_{obs}/dm^3\ mol^{-1}$	Fluorescence enhancement	$K_{obs}/dm^3\ mol^{-1}$	Fluorescence enhancement
139	244 ± 26 (0.99)	3.4	32 ± 3 (0.99)	3.2
147[b]	76 ± 10 (0.98)	1.7	c	c
147[d]	125 ± 11 (0.99)	3.5	8 ± 1 (1.00)	3.5

[a] *The K values were analysed in KaleidaGraph using nonlinear (Levenberg–Marquardt algorithm) curve fitting. The errors reported are the standard errors obtained from the best fit.*
[b] $\lambda_{ex}=299$ nm, $\lambda_{em}=417$ nm.
[c] The K_{obs} and fluorescence enhancement could not be determined because of the small changes in fluorescence.
[d] $\lambda_{ex}=342$ nm, $\lambda_{em}=417$ nm.

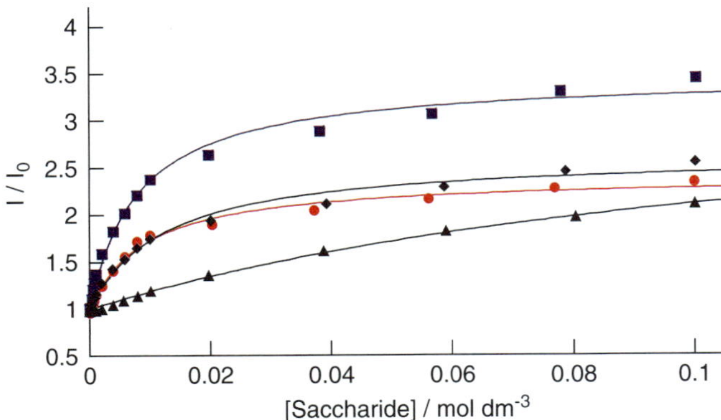

Figure 36 *Relative fluorescence intensity vs. saccharide concentration profile of sensor* **147**$_{(phenanthrene-pyrene)}$ *($2.5 \times 10^{-6}\ mol\ dm^{-3}$) displaying PET at the pyrene fluorophore with (●) D-glucose, (■) D-fructose, (◆) D-galactose, (▲) D-mannose, in 52.1 wt% MeOH pH 8.21 phosphate buffer. $\lambda_{ex}=342$ nm, $\lambda_{em}=417$ nm.*

The fluorophores chosen are all commercially available as their aldehyde derivatives, comparatively inexpensive and have similar photophysical properties.

Sensors **148**$_{(pyrene)}$–**152**$_{(2-naphthalene)}$ all utilise the same general structural motif, this is illustrated in Figure 38. The two phenylboronic acid receptors

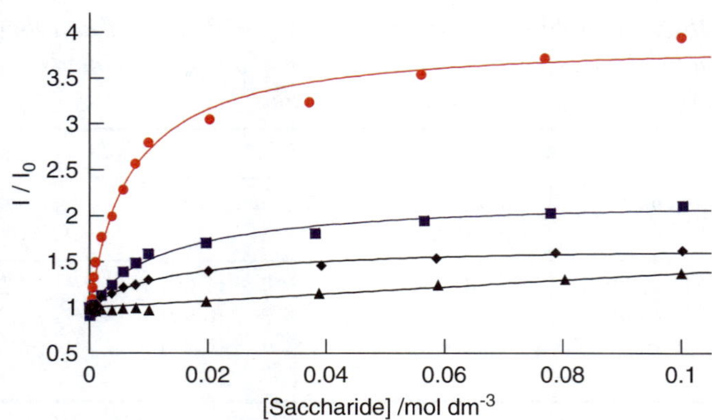

Figure 37 *Relative fluorescence intensity vs. saccharide concentration profile of sensor* **147**$_{(phenanthrene–pyrene)}$ *(2.5×10^{-6} mol dm^{-3}) displaying ET between the phenanthrene and pyrene fluorophores with (●) D-glucose, (■) D-fructose, (◆) D-galactose, (▲) D-mannose, in 52.1 wt% MeOH pH 8.21 phosphate buffer. λ_{ex}=299 nm, λ_{em}=417 nm.*

Key:

Figure 38 *Generic template for the five diboronic acid sensors with variable fluorophore units* **148**$_{(pyrene)}$–**152**$_{(2-naphthalene)}$.

introduce saccharide selectivity while the hexamethylene linker governs the dimensions of the binding cleft, with a proven bias towards D-glucose selectivity.

In order to determine the binding stoichiometry of the complexes formed within the binding sites of the diboronic acid based sensors **148**$_{(pyrene)}$–**152**$_{(2-naphthalene)}$ and elucidate the benefits gained from allosteric interactions, these systems must be contrasted against analogous monoboronic acid reference compounds. These reference compounds **146**$_{(pyrene)}$, **153**$_{(phenanthrene)}$–**156**$_{(2-naphthalene)}$ shall be discussed first (Figure 39).

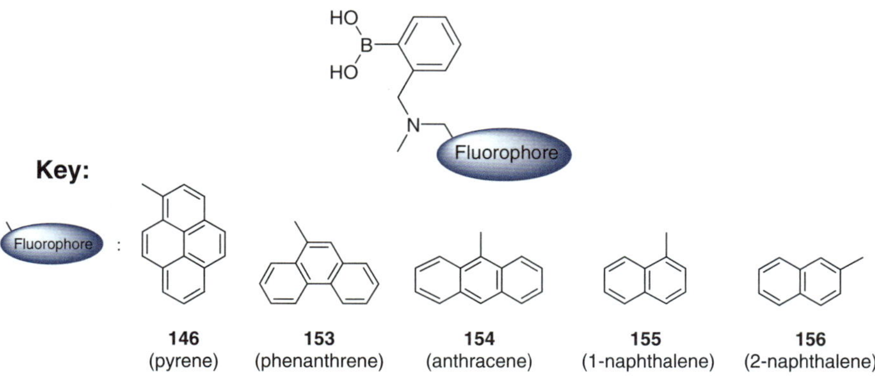

Figure 39   *Generic template for the five monoboronic acid reference compounds with variable fluorophore units* $146_{(pyrene)}$–$156_{(2\text{-}naphthalene)}$.

Table 6   *Fluorescence measurement conditions*

Fluorophore	Concentration/mol dm^{-3}	λ_{ex}/nm	λ_{em}/nm
Pyrene	1.0×10^{-6}	342	397
Phenanthrene	5.0×10^{-6}	299	369
Anthracene	1.0×10^{-6}	370	420
1-Naphthalene	2.5×10^{-6}	275	335
2-Naphthalene	2.5×10^{-6}	274	335

The fluorescence titrations of the monoboronic acid reference compounds $146_{(pyrene)}$, $153_{(phenanthrene)}$–$156_{(2\text{-}naphthalene)}$ and the diboronic acid sensors $148_{(pyrene)}$–$152_{(2\text{-}naphthalene)}$ with D-glucose, D-galactose, D-fructose and D-mannose were carried out in an aqueous methanolic buffer solution [52.1 wt% methanol at pH 8.21 (KCl, 0.01000 mol dm^{-3}; KH$_2$PO$_4$, 0.002752 mol dm^{-3}; Na$_2$HPO$_4$, 0.002757 mol dm^{-3})].[244] The fluorescence intensity of the diboronic acid sensors $148_{(pyrene)}$–$152_{(2\text{-}naphthalene)}$ increased with increasing saccharide concentration. The observed stability constants (K_{obs}) of PET sensors $148_{(pyrene)}$–$152_{(2\text{-}naphthalene)}$ and $146_{(pyrene)}$, $153_{(phenanthrene)}$–$156_{(2\text{-}naphthalene)}$ were calculated by the fitting of emission intensity *vs.* saccharide concentration curves.[258] The emission wavelengths used for each fluorophore in determining the emission intensity are detailed in Table 6. The observed stability constants (K_{obs}) calculated are reported in Tables 7 and 8.

To help visualize the trends of the observed stability constants (K_{obs}) in Tables 7 and 8, the stability constants of the diboronic acid sensors $148_{(pyrene)}$–$152_{(2\text{-}naphthalene)}$ are reported in Figure 40 divided by the stability constants of the analogous monoboronic acid reference compounds $146_{(pyrene)}$, $153_{(phenanthrene)}$–$156_{(2\text{-}naphthalene)}$.

Table 7 *Observed stability constants (K_{obs}) (coefficient of determination, r^2) and fluorescence enhancements for the saccharide complexes of sensors* **146**$_{(pyrene)}$, **153**$_{(phenanthrene)}$–**156**$_{(2\text{-naphthalene})}$ *and* **148**$_{(pyrene)}$–**152**$_{(2\text{-naphthalene})}$

Sensor	D-Glucose		D-Galactose	
	K_{obs}/dm^3 mol^{-1}	Fluorescence enhancement	K_{obs}/dm^3 mol^{-1}	Fluorescence enhancement
148$_{(pyrene)}$	962 ± 70 (0.99)	2.8	657 ± 39 (1.00)	3.1
146$_{(pyrene)}$	44 ± 3 (1.00)	4.5	51 ± 2 (1.00)	4.2
149$_{(phenanthrene)}$	325 ± 58 (0.97)	1.5	611 ± 101 (0.97)	1.4
153$_{(phenanthrene)}$	30 ± 7 (0.98)	1.5	77 ± 12 (0.98)	1.4
150$_{(anthracene)}$	441 ± 76 (0.98)	3.2	536 ± 31 (1.00)	3.1
154$_{(anthracene)}$	61 ± 3 (1.00)	3.4	93 ± 6 (1.00)	3.0
151$_{(1\text{-naphthalene})}$	417 ± 60 (0.98)	6.1	1072 ± 68 (0.99)	5.4
155$_{(1\text{-naphthalene})}$	52 ± 1 (1.00)	5.7	78 ± 5 (1.00)	5.0
152$_{(2\text{-naphthalene})}$	532 ± 57 (0.99)	4.2	894 ± 66 (0.99)	4.1
156$_{(2\text{-naphthalene})}$	35 ± 2 (1.00)	4.5	49 ± 4 (1.00)	4.3

Table 8 *Observed stability constants (K_{obs}) (coefficient of determination, r^2) and fluorescence enhancements for the saccharide complexes of sensors* **146**$_{(pyrene)}$, **153**$_{(phenanthrene)}$–**156**$_{(2\text{-naphthalene})}$ *and* **148**$_{(pyrene)}$–**152**$_{(2\text{-naphthalene})}$

Sensor	D-Fructose		D-Mannose	
	K_{obs}/dm^3 mol^{-1}	Fluorescence enhancement	K_{obs}/dm^3 mol^{-1}	Fluorescence enhancement
148$_{(pyrene)}$	784 ± 44 (1.00)	3.2	74 ± 3 (1.00)	2.8
146$_{(pyrene)}$	395 ± 11 (1.00)	3.6	36 ± 1 (1.00)	3.7
149$_{(phenanthrene)}$	1013 ± 126(0.98)	1.4	134 ± 18 (0.98)	1.4
153$_{(phenanthrene)}$	548 ± 55 (0.99)	1.4	58 ± 8 (0.98)	1.4
150$_{(anthracene)}$	1000 ± 69 (0.99)	3.0	111 ± 6 (1.00)	2.8
154$_{(anthracene)}$	713 ± 35 (1.00)	3.0	61 ± 3 (1.00)	3.0
151$_{(1\text{-naphthalene})}$	964 ± 41 (1.00)	5.5	101 ± 3 (1.00)	5.0
155$_{(1\text{-naphthalene})}$	529 ± 45 (0.99)	5.4	46 ± 1 (1.00)	5.2
152$_{(2\text{-naphthalene})}$	1068 ± 63 (1.00)	3.8	98 ± 4 (1.00)	3.5
156$_{(2\text{-naphthalene})}$	399 ± 34 (0.99)	4.6	40 ± 2 (1.00)	3.8

5.4.1 Inference

It could be thought that altering the appended fluorophore on the modular diboronic acid systems would simply lead to fluorescent sensors with different emission and excitation wavelengths. As can be readily observed from the bar graph in Figure 40 this was not the case.

The relative stabilities clearly illustrate that an increase in selectivity is obtained by cooperative binding through the formation of 1:1 cyclic systems.

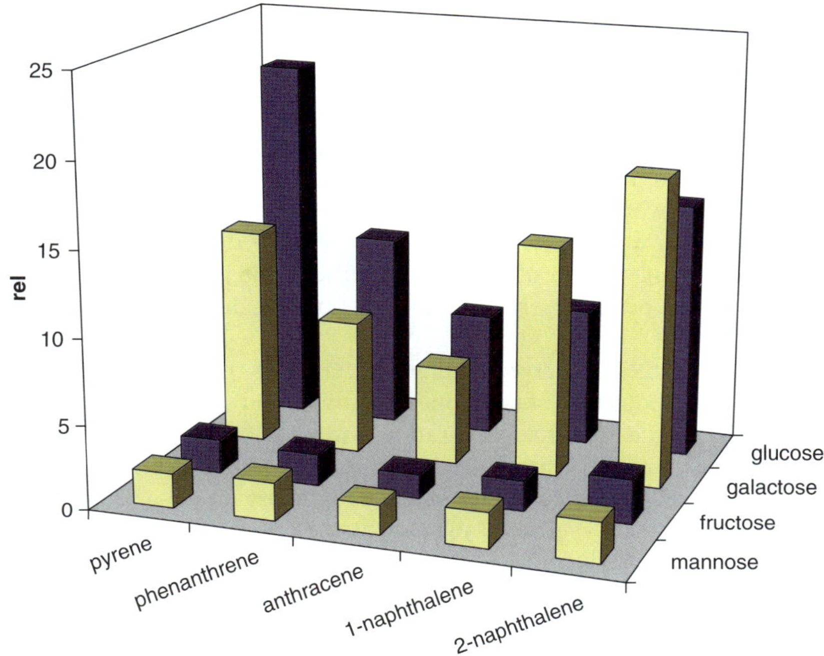

Figure 40 *Observed stability constants (K_{obs}) of the diboronic acid sensors* **148**$_{(pyrene)}$–**152**$_{(2\text{-}naphthalene)}$ *divided by the observed stability constants (K_{obs}) of the corresponding monoboronic acid reference compounds* **146**$_{(pyrene)}$, **153**$_{(phenanthrene)}$–**156**$_{(2\text{-}naphthalene)}$ *to yield relative values with saccharides. The* D-*configuration of the monosaccharides was used throughout this evaluation.*

The large enhancement of the relative stability observed for the 1:1 cyclic systems (formed with D-glucose and D-galactose) are clearly contrasted with the small two-fold enhancement observed for the 2:1 acyclic systems (formed with D-fructose and D-mannose).

The results in Figure 40 indicate that the best match between receptor and guest is for sensors **148**$_{(pyrene)}$, **153**$_{(phenanthrene)}$ and **150**$_{(anthracene)}$ with D-glucose and for sensors **151**$_{(1\text{-}naphthalene)}$ and **152**$_{(2\text{-}naphthalene)}$ with D-galactose.

The bar graph in Figure 40 describes a saddle-like trend from pyrene through to 2-naphthalene with a change in selectivity from D-glucose to D-galactose occurring in the middle of the series. To account for such a trend these results must be a function of more than one variable.

5.4.1.1 Solvation. The first factor to be considered was the size of the fluorophore's π-surface. From Figure 40 it can be observed that the largest relative value was for sensor **148**$_{(pyrene)}$ with D-glucose. Pyrene, with four fused benzene rings, represents the largest fluorophore in the series and as such endows sensor **148**$_{(pyrene)}$ with the largest planar aromatic π-surface of any of the sensors **148**$_{(pyrene)}$–**152**$_{(2\text{-}naphthalene)}$. As the size of the fluorophore is reduced to three fused benzene rings in phenanthrene and anthracene the relative values are seen to diminish. As the fluorophore size decreases further still to the two fused benzene rings in 1- and 2-naphthalene there is a distinct change in the direction of the trend, the relative values increase, with a concomitant change in selectivity from D-glucose to D-galactose. Given the proximity of the fluorophore's planar aromatic π-surface to the binding pocket it seemed reasonable to infer that the hydrophobicity within the binding pocket would be a function of the size of the fluorophore. If so sensor **148**(pyrene) would therefore have the most hydrophobic environment within the binding pocket of any of the sensors **148**(pyrene)–**152**(2-naphthalene), this environment being observed to have the best D-glucose selectivity across the series.

To rationalise these observed changes in terms of the complementarity between the hydrophobicities of the monosaccharides and sensors **151**(1-naphthalene)–**152**(2-naphthalene) the reported hydration values for the monosaccharides were examined. From the literature it appears that establishing the hydration characteristics of monosaccharides such as D-glucose and D-galactose is non-trivial, however, there is general agreement that D-galactose is slightly more hydrophobic than D-glucose.[259–261] The opposite of what we would expect based on the above observations.

The interactions between aromatic hydrocarbons and monosaccharides in aqueous solutions have been determined. Janado *et al.* documented that aqueous solutions of D-galactose were found to dissolve more benzene, naphthalene and biphenyl than the equivalent aqueous solutions of D-glucose.[262,263] A result consistent with D-galactose being the more hydrophobic of the two saccharides.

In attempting to establish a rationale for the values displayed in Figure 40 a number of factors have to be considered: the size, orientation and proximity of the fluorophores in sensors **148**(pyrene)–**152**(2-naphthalene) to the binding pocket; the hydrophobicities of the monosaccharides in their relative compositions and the manner in which complexation with boronic acids will change this (as discussed above D-glucose will predominantly favour the α-D-glucofuranose form and D-galactose will favour an α-D-galactofuranose or α-D-galactopyranose (twist-boat) form)[43,235] as well as the covalent interactions of the saccharide hydroxyl groups with the diboronic acid receptors on binding. Rationalising these trends in a quantitative manner would therefore require a considerable computational effort given the complex and dynamic nature of these systems, however, in considering these experimental observations a number of important features come to light.

Since Emil Fischer's seminal article of 1894 in which the hydrated polar groups within yeast's binding site were described as a locked gate that could only be opened by the key polar groups of α-glucosides (and not β-glucosides) much work has been done on understanding the cause of such well-defined stereospecific recognition.[264,265] While the physical fit of the "key" within the "lock" could be intuitively understood, further research demonstrated that selectivity was not just a facet of enthalpic stabilisation but relied intimately on the increase in entropy that occurred when water was displaced from the binding site. In turn this entropic stabilisation had to be played off against the decrease in entropy that occurred when the perturbed solvent shell surrounding the saccharide was displaced.

For most of the hexoses and pentoses there is little overall difference in hydrophobicity, nevertheless, it is the case that very slight differences in the stereochemical configuration of monosaccharides' hydroxyl groups can lead to significant changes in solvation properties. One of the most significant deviations from the general trend within the hexoses is the presence of an axial C4 hydroxyl group.[260,266] This facet has been documented to cause D-galactose to have a substantially diminished fit in water compared to D-glucose and as such the two monosaccharides have been categorised as having markedly different hydration properties (*i.e.* while their observed hydrophobicities may appear to be similar, the degree of disorder they introduce to the water molecules surrounding them is quite different.)[259,261,267]

Water molecules are therefore crucial in controlling the selective recognition of saccharides within aqueous systems and it is the case that the restructuring of perturbed surface water provides an important force in controlling these molecular associations.[268] Therefore if correctly approached the solvation effects observed within our simple diboronic acid sensors may provide the impetus for a more refined discrimination to be made between the molecular recognition of D-glucose and D-galactose in aqueous systems through careful control of the hydrophobicity of the binding pocket.

This observation may prove of particular interest not only in sensor design but also in regard to the significant work being undertaken on saccharide transport through organic membranes (such as lipid bilayers) *via* boronic acid carriers. In these systems D-galactose is noticeable by its absence as, to date, no data has been reported regarding the transport of D-galactose through hydrophobic membranes using boronic acid carriers.[269–274]

5.4.1.2 Steric Crowding. If the size of the fluorophore's π surface was the only factor to be considered, little difference would be expected between the results observed for sensors **151**(1-naphthalene) and **152**(2-naphthalene). This was not the case. In comparing sensors **151**(1-naphthalene) and **152**(2-naphthalene) there was no difference in size between the fluorophore's hydrophobic π-surfaces, only a difference in connectivity and therefore the relative alignment of the fluorophores with regard to the binding cleft. In explaining the difference between the relative values, the number of *peri*-hydrogens on the fluorophores was considered. The fluorophore with the highest number of *peri*-hydrogens is anthracene with two, then pyrene, phenanthrene and 1-naphthalene with one each and then 2-naphthalene with none, this is illustrated in Figure 41.

This ties in well with the observed results in the bar graph of Figure 40. Anthracene with the greatest number of *peri*-hydrogens displays the lowest

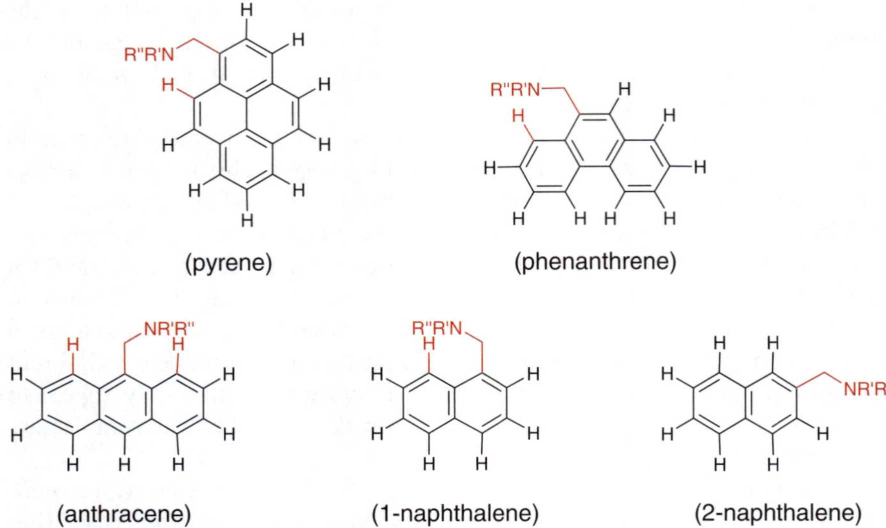

Figure 41   *The five fluorophores used for the mono- and di-boronic acid systems. For clarity the methylamine fragment, peri-hydrogen and their bonds are highlighted in red. The two groups are illustrated here in their eclipsed conformation.*

relative stability. These *peri*-hydrogens can be considered to increase the steric crowding within the binding pocket and as such this hypothesis seems quite reasonable. It also fits the increase in sensitivity observed for 2-naphthalene over 1-naphthalene.

5.4.1.3 Denouement. These results demonstrate that in a fluorescent PET saccharide sensor with two phenylboronic acid groups, a hexamethylene linker and a fluorophore, the choice of the fluorophore is crucial. The fluorophore not only defines the emission wavelength but also influences the environment within the binding site with the overall selectivity being fluorophore dependent.

An informed decision will therefore have to be taken in the future design of fluorescent sensors, such that the polarity of the chosen guest species complements the solvation within the binding pocket. While not a direct premise of complementarity between hydrophobicity of the appended fluorophore of the sensor and the pyranose form of the guest monosaccharide, it appears to be the case that boronic acids display enhanced selectivity for D-glucose over D-galactose when the hydrophobicity of the binding pocket is increased.

In addition to considering solubility, a minimisation of the *peri*-hydrogens should reduce steric crowding within the binding pocket and, as demonstrated with D-glucose and D-galactose, increase the relative stability of complexes formed with diboronic acid sensors.

5.5   Other Approaches

Following publication of the modular construct used by our research group in the design of boronic acid based fluorescent PET sensors in 2001,[242] two research groups have published further data on libraries of sensors assembled in a modular fashion.

5.5.1   Wang and Co-workers

Recently Wang and co-workers have documented a range of diboronic acid sensors Figure 42 (**a**)–(**z**).[251,275,276] It can be seen by examining the generic template used that the sensors are designed around the known core of sensor **80**, the first diboronic acid sensor to display selectivity for D-glucose. In this construct the number of carbon atoms from one *N*-methyl-*O*-(aminome-thyl)phenylboronic acid nitrogen atom to the other is increased substantially. Six carbon atoms separate each of the adjacent amine – amide nitrogen atoms with anthracene cores rigidifying this section of the molecule and introducing possible interactions through either π–π stacking or steric encumbrance. The variable linkers examined augment the length of these rigid linkers further still.

Figure 42 (**f**) with the *para*-benzene linker was found to be selective for sialyl Lewis X.[251,276] Figure 42 (**g**) with the *ortho*-xylene linker was found to be selective for D-glucose.[275] In replacing the *ortho*-xylene linker of Figure 42 (**g**) with the flexible butyl linker of Figure 42 (**b**) the number of carbon atoms in the linker remained the same but the structural rigidity of the linker was lost, this led to a halving of the observed stability constant (K_{obs}). In reintroducing the rigidity but changing the geometry and spacing of the core unit to the *ortho*-benzene linker Figure 42 (**h**) the observed stability constant (K_{obs}) was seen to decrease further still.

sialyl Lewis X

The observed stability constants (K_{obs}) for the selected sensors in Figure 42 were: (**g**), 34 M^{-1} with D-fructose, 1 470 M^{-1} with D-glucose and 30 M^{-1} with D-galactose; (**b**), 80 M^{-1} with D-fructose, 640 M^{-1} with D-glucose and 110 M^{-1} with D-galactose; (**h**), 280 M^{-1} with D-fructose, 180 M^{-1} with D-glucose and 30 M^{-1} with D-galactose in a solution of 1:1 (v/v) methanol/0.1 M aqueous phosphate buffer solution at pH 7.4. The observed stability constant (K_{obs}) for sensor Figure 42 (**f**) with sialyl Lewis X was not reported.

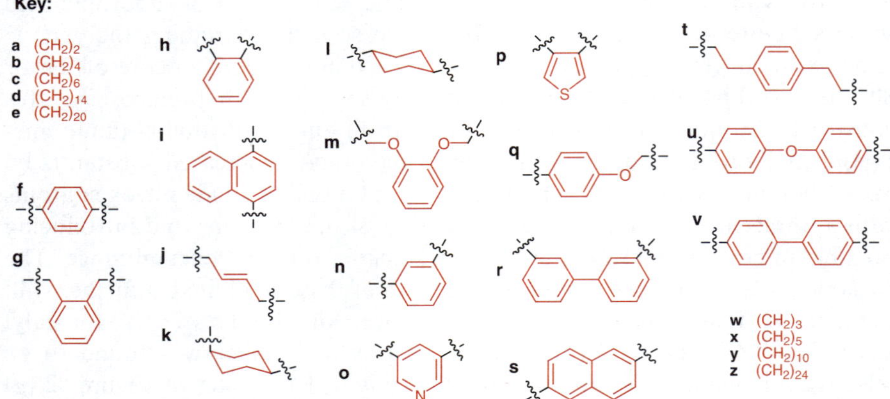

(f)

Key:

a $(CH_2)_2$
b $(CH_2)_4$
c $(CH_2)_6$
d $(CH_2)_{14}$
e $(CH_2)_{20}$

w $(CH_2)_3$
x $(CH_2)_5$
y $(CH_2)_{10}$
z $(CH_2)_{24}$

Figure 42 *Range of diboronic acid sensors (**a**)–(**z**) with the inter-changeable linker fragments highlighted in red; (**f**) displayed selectivity for sialyl Lewis X and (**g**) displayed selectivity for D-glucose.*

5.5.2 Hall and Co-workers

In an article published in 2004, Hall and co-workers documented the first parallel, solid phase synthesis of modular boronic acid based sensors.[277] The series was developed from a range of common components allowing the rapid assembly of a library of compounds, with the use of semi-preparative high performance liquid chromatography (HPLC) to ensure sensors of satisfactory purity. This approach allowed the structures of the inter-amine linkers to be altered (selectivity was once again found for D-glucose with a linker six carbon atoms in length), however, the investigation went on to assess the potential role of a third boronic acid receptor moiety in the recognition of disaccharides (Figure 43) and the effect of introducing unencumbering electron-withdrawing and electron-donating groups *para* to the arylboronic acid (Figure 44).

The triboronic acid sensors were titrated against four disaccharides: lactulose, melibiose, turanose and trehalose. Across the range of sensors and guests examined no benefit was found from three (*vs.* two) boronic acid receptor units. For example, the observed stability constant (K_{obs}) of triboronic acid Figure 43 (**1c, 2a**) with lactulose was 200 M^{-1}, this value can be contrasted with the value obtained for the analogous diboronic acid (**2a** derivative) that displayed an

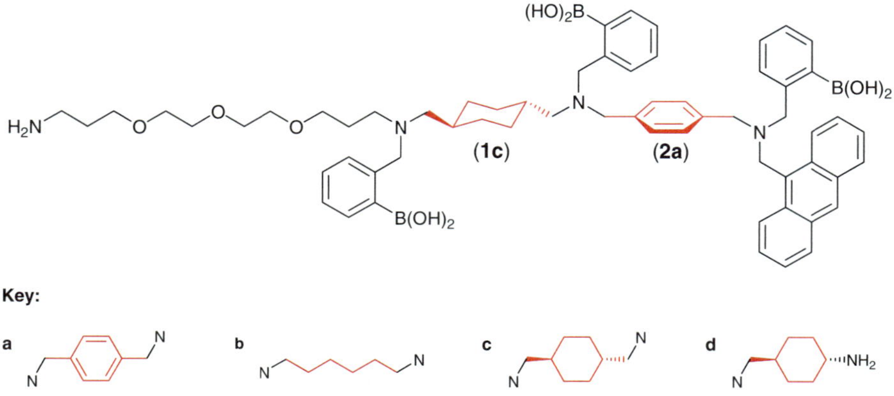

Figure 43 *Triboronic acid sensors assembled via combinatorial synthesis to evaluate the potential of increased allosteric binding effects through three point binding (the inter-changeable linker units are highlighted in red).*

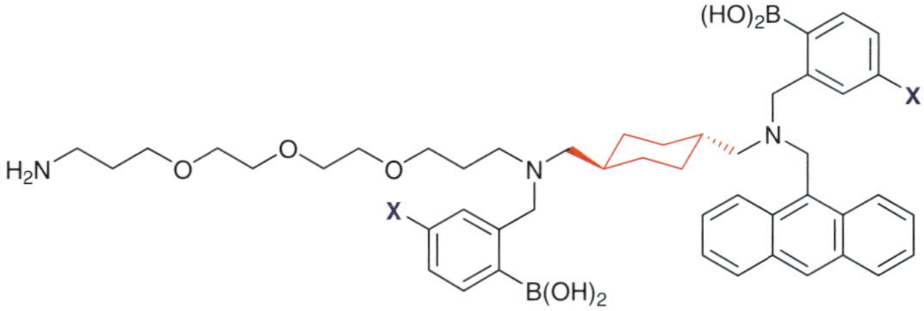

Figure 44 *Diboronic acid modular systems with an inter-changeable linker section (highlighted in red) and inter-changeable electron-withdrawing and electron donating groups para to the arylboronic acid (highlighted in blue).*

observed stability constant (K_{obs}) of 220 M^{-1} with lactulose in 0.010 mol dm^{-3} phosphate buffer at pH 7.8 in a 1:1 water/methanol mixture.

Building on these observations Hall and co-workers examined the dependence of complexation on the electronic characteristics of the arylboronic acid receptors. By altering the Lewis acidity at boron it was believed that two main features in the molecular recognition event could be altered: the strength of the binding interaction with the saccharides and the strength of the N–B interaction controlling the fluorescence intensity. This hypothesis was tested by introducing electron-withdrawing and electron-donating groups at the *para* position of the arylboronic acid ring.

Five substituent groups were considered in all: methoxy, fluoro, methoxycarbonyl, cyano and nitro, with *para*-substituent parameters, σ_p: −0.12, 0.15, 0.44, 0.70 and 0.81, respectively.[278] If Figure 44 is considered and lactulose is used as the model disaccharide, the binding measurements display a qualitative trend that electron poor phenylboronic acids are preferable for binding. This observation was rationalised on the basis that on increasing the Lewis acidity at boron the N–B interaction becomes stronger developing more of a tetrahedral

Table 9 *Observed stability constants (K_{obs}) for the general structure illustrated in Figure 44 with lactulose, determined in 0.010 mol dm^{-3} phosphate buffer at pH 7.8 in a 1:1 water/methanol mixture*

X	Linker	σ_p	$(K_{obs})/dm^3\ mol^{-1}$
OMe	**(c)**	−0.12	150
F	**(c)**	0.15	585
CN	**(c)**	0.70	1020
CN	**(b)**	0.70	1870

character at boron. This in turn reduces the ring strain in the developing boronic ester. In addition to this, acidifying boron and reducing its pK_a also provides the substantial benefit of allowing the sensor to function at lower pH (Table 9).

In enhancing the observed stability constants (K_{obs}) the use of methoxycarbonyl, cyano and nitro groups appeared to be particularly effective. Overall the largest observed stability constant (K_{obs}) reported with lactulose was with the diboronic acid illustrated in Figure 44 with a hexamethylene linker (**b**) and a *para*-cyano electron withdrawing group.[277]

5.6 Summary

- Optimum D-glucose selectivity is obtained across the range of modular diboronic acid sensors **140**$_{(n=3)}$–**145**$_{(n=8)}$ when the two amine nitrogens are separated by a linker six carbon atoms in length.
- The change in selectivity from D-glucose to D-galactose as a function of the increasing dimensions of the binding pocket can be ascribed to the relative availability of *syn*-periplanar hydroxyl pairs and the stability of D-glucose in its furanose form and D-galactose in a twist boat conformation of its pyranose form.
- Where significant binding was observed between diboronic acids and a disaccharide, the disaccharide possessed a terminal residue that could inter-convert between its pyranose and furanose forms.
- Where the above condition was met steric crowding at the secondary binding site was thought to reduce the strength of the complex formed between the diboronic acid and disaccharide, compared to the complex formed between the same diboronic acid and the analogous monosaccharide.

Intramolecular energy transfer between suitable fluorophores provides a highly efficient way of discriminating between saccharides based on the binding motif of the complex formed.

Other Types of Sensor

Lucky is he who can understand the causes of things

Virgil (70–19 B.C.)

6.1 Colorimetric Sensors

Colorimetric sensors for saccharides are of particular interest in a practical sense. If a system with a large colour change can be developed, it could be incorporated into a diagnostic test paper for D-glucose, similar to universal indicator paper for pH. Such a system would make it possible to measure D-glucose concentrations without the need of specialist instrumentation. This would be of particular benefit to diabetics in developing countries.

157

158

A fairly recent development has been the study of the effect of saccharides on the colour of dyes containing boronic acid functionality. Boronic acid azo dyes have been known for over 40 years having been used for investigations in the treatment of cancer by a technique called boron neutron capture therapy (BNCT).[279,280] It was not until the 1990s that related dyes and their interaction with saccharides were studied. Russell has synthesised a boronic acid azo dye from *m*-aminophenylboronic acid, which was found to be sensitive to a selection of saccharides.[281] The usefulness of the dye as a D-glucose-monitoring agent in fermentation processes was proven by tests in beef-broth solution (used in the growth of bacteria) containing various proteins, lipids and salts. Nagasaki observed that chromophores containing boronic acid moieties **157** and **158** (which aggregate in water) changed colour and deaggregated upon complexation with saccharides.[282] This was rationalised by the boronic acid–saccharide complexation increasing the hydrophilicity of the bound species.

159

Takeuchi has prepared a boronic acid dye **159**, which undergoes an absorption spectral change on addition of nucleosides.[283] The boronic acid binds with the ribose and the dimethylaminophenylazo moiety can stack with the adenine of the nucleoside.

160

The ICT sensor **160** prepared by Sandanayake, employs an intramolecular interaction between the tertiary amine and the boronic acid group to promote colour changes on addition of saccharides.[284] The electron-rich amine creates a basic environment around the electron-deficient boron centre, which has the effect of inducing the boronic acid–saccharide interaction and reducing the working pH of the sensor. Electronic changes associated with this decrease in the pK_a of the boronic acid moiety on saccharide complexation, are transmitted to the neighbouring amine. This creates a spectral change in the connected-ICT chromophore, which can be detected as a change in colour. The pK_a related to the boron–nitrogen interaction of **160** shifts on the addition of saccharides. The largest pK_a shift was found for D-fructose ($\Delta pK_a = 3.31$) and the smallest for simple diols such as ethylene glycol ($\Delta pK_a \approx 0$). The observed stability constant (K_{obs}) for **160** was 138 M^{-1} for D-fructose in water at pH 7.6. The stability constant with D-glucose was not determined due to small spectral changes.

161

Shinmori has synthesised a diboronic acid–saccharide receptor bearing a photoresponsive azobenzene group **161**, which was used as a light-gated saccharide sensor.[285] When the azobenzene unit is switched by photoirradiation from the more stable *trans*-conformation to the thermodynamically unfavourable *cis*-isomer, it shows high D-glucose and D-allose selectivity. The formation of cyclic 1:1 complexes between saccharide and the dye in its *cis*-geometry explains the selectivity order.

162

163

Koumoto demonstrated that azobenzene derivatives bearing one or two aminomethylphenylboronic acid groups **162** and **163** can be used for practical colorimetric saccharide sensing in "neutral" aqueous media.[286] The observed stability constants (K_{obs}) for **163** were 433 M^{-1} for D-fructose and 13.0 M^{-1} for D-glucose in 1:1 (v/v) methanol/water at pH 7.5 (phosphate buffer).

164

Koumoto has cleverly used the boronic acid–amine interaction to the molecular design of an intermolecular sensing system for saccharides.[287] 3-Nitrophenylboronic acid interacts with the pyridine nitrogen of 4-(4-dimethylaminophenylazo)-pyridine **164** in methanol and changes its colour from yellow to orange. Added saccharides form complexes with the boronic acid and enhance the acidity of the boronic acid group. As a result the boron–nitrogen interaction becomes stronger and the intensified intramolecular charge-transfer band changes the solution colour to red.

165: X = *p*-NO$_2$
166: X = *p*-SO$_3$H
167: X = *p*-CO$_2$H
168: X = *p*-OMe
169: X = *m*-CO$_2$H

170

James and co-workers recently prepared a diazo dye system **165**, which shows a large visible colour change from purple to red on saccharide binding.[288,289] With azo dye **165** the wavelength maximum shifts by approximately 55 nm to a shorter wavelength upon saccharide complexation. The observed stability constants (K_{obs}) for **165** were 2550 M^{-1} for D-fructose and 123 M^{-1} D-glucose in 52.1 wt% methanol/water at pH 11.32 (carbonate buffer).

With dye molecule **160**, it was proposed that at intermediate pH a boron–nitrogen interaction exists, whereas at high and low pH this interaction is broken. What makes the equilibria of dye molecule **165** more interesting is the presence of the *anilinic hydrogen*, which can give rise to different species at high pH.

In the absence of saccharide, at pH 11.32, the observed colour of **165** is purple and in the presence of saccharide the colour is red. In the presence of saccharide, the N–B interaction becomes stronger. The increased N–B interaction causes the N–H proton to become more acidic. Therefore at pH 11.32, the saccharide–boronate complex dehydrates (loss of H+ from aniline and OH− from boronate) to produce a red species with a covalent N–B bond.

These equilibrium species explain why dye molecule **160** did not give a visible spectral shift on saccharide binding. With **160** there is no possibility of dehydration, so a strong boron–nitrogen bond cannot be formed, hence no spectral shift is observed. This hypothesis has been confirmed by evaluating **170**, which does not have an anilinic hydrogen. No colour change was observed for **170** on addition of saccharides at pH 11.32.[289] James and co-workers have also carried out a detailed investigation on a series of azo dyes with both electron-donating and -withdrawing groups **165**, **166**, **167**, **168** and **169** and determined that a strong electron-withdrawing group is required to produce a colour change.[289]

171

172

Lakowicz has also prepared boronic acid azo dye molecules **171** and **172** in which direct conjugation with the boron centre is possible. In particular, the azo dye **172** produces a visible colour change from yellow to orange at pH 7.[290] The observed stability constants (K_{obs}) for **172** were 158 M^{-1} for D-fructose and 2.7 M^{-1} for D-glucose in water at pH 7.0 (phosphate buffer).

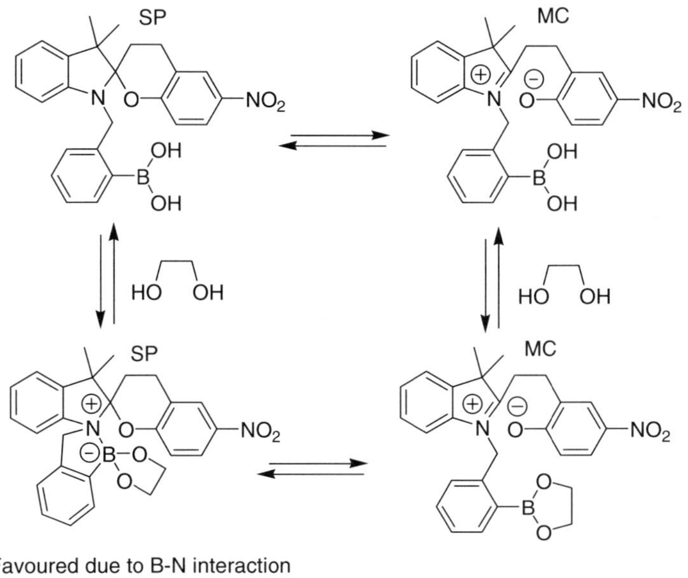

Scheme 42 *The effect of saccharides on the SP vs MC equilibrium.*

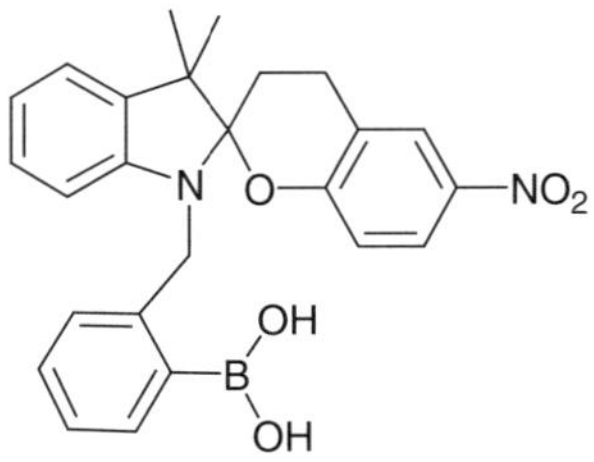

173

Shinmori has shown that a boronic acid-appended spirobenzopyran **173** undergo changes in the absorption spectra on the addition of saccharides.[291] The added saccharides change the position of the merocyanine (MC) to spiropyran (SP) equilibrium and hence change the colour of the system. With added saccharide, the SP structure is favoured due to a stronger N–B interaction in the saccharide complex (Scheme 42).

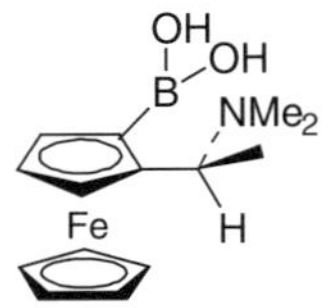

174 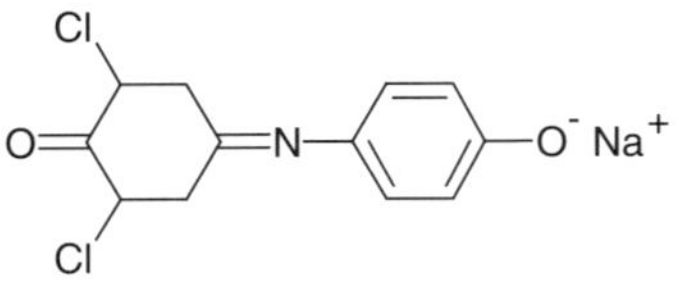 **175**

Yamamoto has used the change in redox potential of **174** to modulate the colour of **175**. When saccharides bind with **174**, the pK_a of the boronic acid is lowered this in turn alters the redox potential of the ferrocene moiety. Dye molecule **175** was chosen since its redox potential was such that reduction by the ferrocence unit becomes facile as the redox potential of the ferrocene lowers on addition of saccharides.[292]

176 (*R*)

177 (*R*)

Mizuno has investigated the use of chiral salen cobalt(II) complexes **176** (*R*) and **177** (*R*).[293] Spectroscopic changes in the metal complexes were used to monitor the formation of the saccharide complexes. Chiral discrimination was observed with **176** (*R*), which showed two-fold selectivity for L-allose over D-allose. Mizuno has also used a prochiral salen cobalt(II) complex, the binding and chirality was monitored using circular dichroism (CD) spectroscopy.[294] The observed stability constants (K_{obs}) for **176** (*R*) were 2760 M^{-1} for D-fructose, 2700 for L-fructose, 240 M^{-1} for D-glucose, 250 M^{-1} for L-glucose, 360 M^{-1} for D-allose and 780 M^{-1} for L-allose in 1:1 (v/v) methanol/water at pH 9.5 (carbonate buffer). The observed stability constants (K_{obs}) for **177** (*R*) were 170 M^{-1} for D-glucose, 210 M^{-1} for L-glucose, 200 M^{-1} for D-allose and 320 M^{-1} for L-allose in 1:1 (v/v) methanol/water at pH 9.5 (carbonate buffer).

178

Yam and Kai[295] and a detailed re-investigation by Mizuno[296] have explored the sensing properties of a boronic acid appended rhenium(I) complex **178**. This system and **176** and **177** illustrate how metal chelation can be used to extend the working wavelength of a sensor.

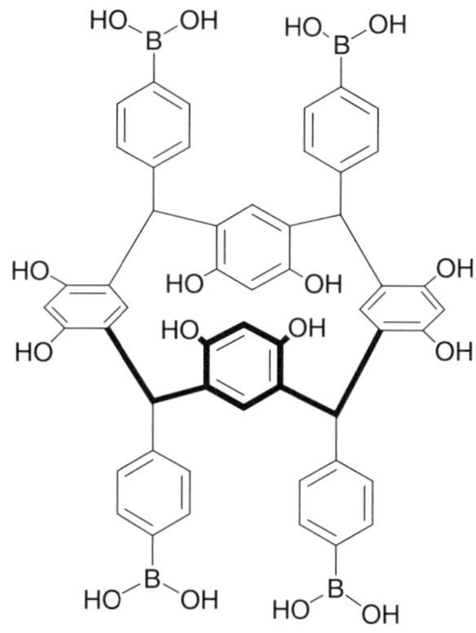

179

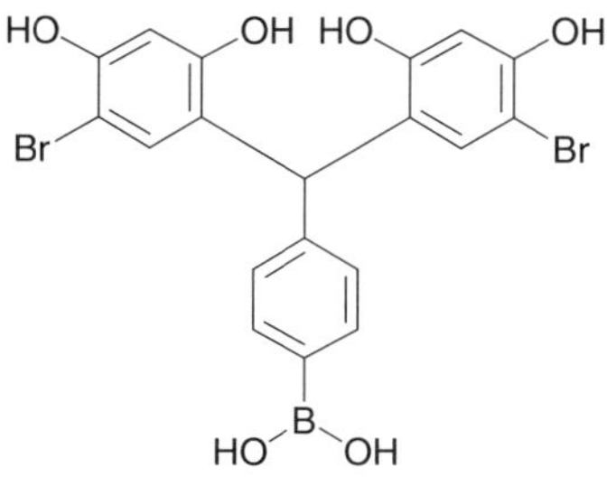

180

181

Strongin has prepared a tetraboronic acid resorcinarene system **179** for the visual sensing of saccharides.[297,298] Characteristic colour changes were observed for specific saccharides, D-glucose phosphates and amino sugars on gentle heating in DMSO. Further, work by Strongin with another resorcinol derivative, **180**, has shown that oxygen promotes the colour changes and that the resorcinol hydroxyl groups play a key role in the colour formation of the solutions.[300] The mechanism of colour change points to xanthenes as *in situ* chromophores formed by heating resorcinols in DMSO. Non-boronic acid receptors also produce coloured solutions but to a lesser extent. In these cases, the colour is due to hydrogen bonding between aldonic acids (heating sugars in DMSO produces aldonic acid derivatives) and the hydroxyls of the *in situ* xanthene chromophore.[299,300]

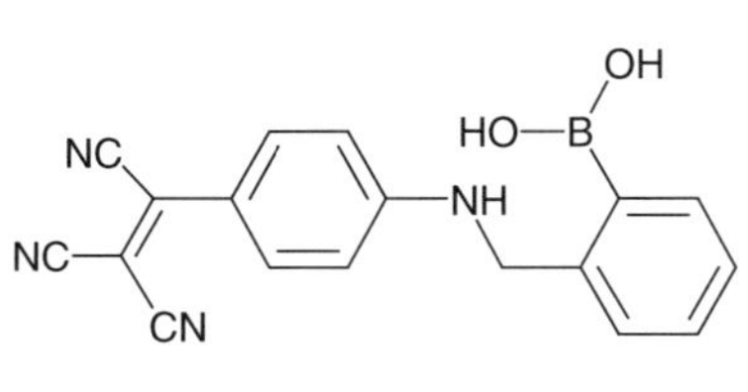

The results obtained with **165, 166, 167, 168** and **169** led James and co-workers to prepare the strongly electron-withdrawing tricyanovinyl dye **181**. The pK_a of **181** (7.81) was much less than **165** (10.2), resulting in a visible colour change on addition of saccharides at a much lower pH (8.21).[301] The observed stability constants (K_{obs}) for **181** were 170 M^{-1} for D-fructose and 8.3 M^{-1} for D-glucose in 52.1 wt% methanol/water at pH 8.21 (phosphate buffer).

182   **183**

Sato has prepared stilbazolium boronic acids **182** and **183** and demonstrated the suitability of this unit for the optical sensing of saccharides.[302] The observed stability constants (K_{obs}) for **182** and **183** were 220 and 280 M^{-1} for D-fructose and 4 and 6 M^{-1} D-glucose in 4% acetonitrile/water at pH 7.0 (phosphate buffer).

184   **185**

Wang has prepared nitrophenol boronic acids **184** and **185**, which show large shifts in the UV on addition of saccharides. The changes have been attributed to a change in the balance of the phenolate to boronate equilibria in the presence of saccharides.[303] The observed stability constants (K_{obs}) for **184** and **185** were 245 and 13.5 M^{-1} for D-fructose and 8.0 and 1.2 M^{-1} for D-glucose in 4% (v/v) methanol/water at pH 7.4 (phosphate buffer).

6.2   Electrochemical Sensors

Electrochemical detection of saccharides by enzymatic decomposition of saccharides is the basis of most current commercial D-glucose biosensors.[304] The development of boronic acid-based electroactive saccharide receptors for D-glucose is also possible. However, the main value of the boronic acid-based synthetic systems is that they could provide selectivity for a range of saccharides other than D-glucose.

Chiral ferroceneboronic acid derivatives (− or +)-**174** have been synthesised by Ori and tested for chiral electrochemical detection of monosaccharides.[305] The best discrimination was observed for L-sorbitol and L-iditol at pH 7.0 in 0.1 mol dm^{-3} phosphate buffer solution.

Moore and Wayner have explored the redox switching of saccharide binding with commercial ferrocene boronic acid.[306] From their detailed investigations, they have determined that binding constants of saccharides with the ferrocenium form are about two orders of magnitude greater than for the ferrocene form. The increased stability is ascribed to the lower pK_a of the ferrocenium (5.8) than ferrocene (10.8) boronic acid.

186

Fabre has investigated the electrochemical sensing properties of the boronic acid substituted bipyridine iron(II) complex **186**.[307] On addition of 10 mM D-fructose, the oxidation peak was shifted by 50 mV towards more positive values.

187   **188**

James and co-workers have prepared a ferrocene monoboronic acid **187** and diboronic acid **188** as electrochemical saccharide sensors.[308] The monoboronic acid system **187** has also been prepared and proposed as an electrochemical sensor for saccharides by Norrild.[309] The electrochemical saccharide sensor **188** contains two boronic acid units (saccharide selectivity), one ferrocene unit (electrochemical read out) and a hexamethylene linker unit (for D-glucose selectivity). The electrochemical sensor **188** displays enhanced D-glucose (40 times) and D-galactose (17 times) selectivity when compared to the monoboronic acid **187**.

6.3 Assay Systems

So far, we have discussed the development of integrated molecular sensors using boronic acids. The systems contain a receptor and reporter (fluorophore or chromophore) as part of a discrete molecular unit. However, another approach towards boronic acid-based sensors is also possible where the receptor and a reporter unit are separate as in a competitive assay. A competitive assay requires that the receptor and reporter (typically a commercial dye) associate under the measurement conditions. The receptor–reporter complex is then selectively dissociated by the addition of the appropriate guests. When the

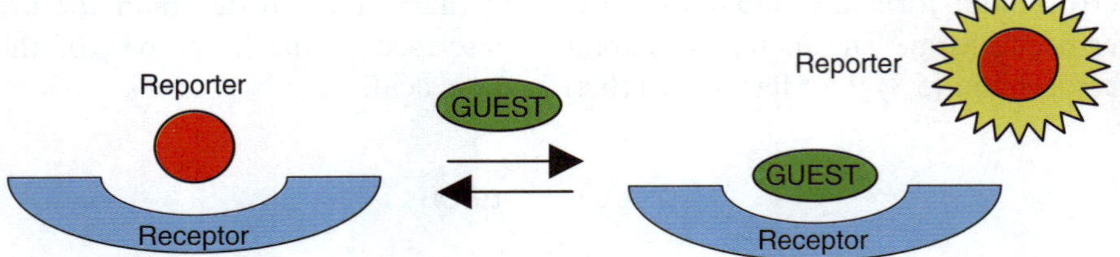

Scheme 43   *Cartoon depicting the function of an assay system.*

reporter dissociates from the receptor, a measurable response is produced (Scheme 43).

The competitive assay approach to novel chemosensors has been pioneered by Anslyn.[310–313] These competitive systems are particularly interesting because they reduce the synthetic complexity of the receptor.

189

5-carboxyfluorescein

190

pyrocatechol violet

Anslyn has recently reported two very elegant systems based on boronic acid receptors. Although, the Anslyn system involves a competitive colourimetric assay, there is no reason why the system cannot be extended to a fluorimetric assay through the choice of appropriate dye molecules. The C_3 symmetric tripodal boronic acid **189** is a selective receptor for D-glucose-6-phosphate.[314] The binding of D-glucose-6-phosphate is measured through the competitive displacement of 5-carboxyfluorescein. Visually, addition of the D-glucose-6-phosphate caused a decrease in the absorption of light at 494 nm allowing the concentration of the guest to be monitored directly within the visible spectrum. The observed stability constants (K_{obs}) for receptor **189** were 3×10^2 M^{-1} with 5-carboxyfluorescin and 1.6×10^3 M^{-1} with D-glucose-6-phosphate in an aqueous 70 wt% methanol in aqueous HEPES buffer at pH 7.4.

Anslyn has also prepared more elaborate C_3 symmetric tri-podal boronic acid receptors. The binding of heparin and **190**, is monitored through displacement of pyrocatechol violet.[246] The observed stability constant (K_{obs}) for **190** was 3.8×10^4 M^{-1} for unfractionated heparin (UFH) in water at pH 7.4 (HEPES buffer).

191

Alizarin complexone

192

bromopyrogallol red

With sensor **191** the binding of the tartrate or malate anions can be detected through the competitive displacement of alizarin complexone.[315] The same sensor system was used for the analysis of malate in pinot noir grapes.[316] When **192** was paired with pyrocatechol violet an assay suitable for the detection of gallic acid in Scotch whiskies was developed. An increase in the concentration of gallic acid correlated with the age of the whiskies.[317] A combination of **191** and **192** and two indicators pyrocatechol violet and bromopyrogallol red can be used to detect the concentrations of tartrate and malate in mixtures.[318] Using

192 and pyrocatechol violet, the reaction kinetics for the formation of tartaric acid by the dihydroxylation of malic acid could be followed.[319]

Anslyn has also elegantly paired chiral boronic acids with a variety of indicators to develop enantioselective assays for α-hydroxyl carboxylates and diols.[320,321]

193: *ortho*
194: *meta*

1,5-ADS   2,6-ADS

Arimori has used compounds **193** and **194** with 1,5- or 2,6-anthraquinone disulfonates (ADS) in a competitive system for the fluorescence detection of D-fructose.[322] 1,5- or 2,6-ADS binds with **193** or **194** and quenches the fluorescence, addition of D-fructose causes decomplexation and fluorescence recovery.

Lakowicz has also used competitive interactions between a ruthenium metal–ligand complex, a boronic acid derivative and D-glucose.[323] The metal–ligand complex forms a reversible complex with 2-tolylboronic acid or 2-methoxyphenylboronic acid. Complexation is accompanied by a several-fold increase in the luminescent intensity of the ruthenium complex. Addition of D-glucose results in decreased luminescent intensity, which appears to be the result of decreased binding between the metal–ligand complex and the boronic acid. Ruthenium metal–ligand complexes are convenient for optical sensing because their long luminescent decay times allow lifetime-based sensing with simple instrumentation. However, with this system 40 mM D-glucose produces only an 11% change in intensity.

195

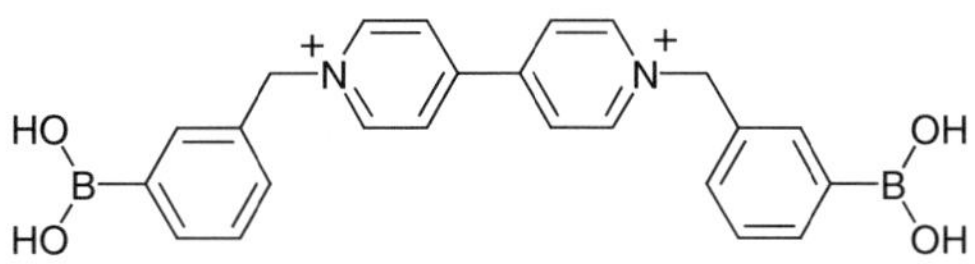

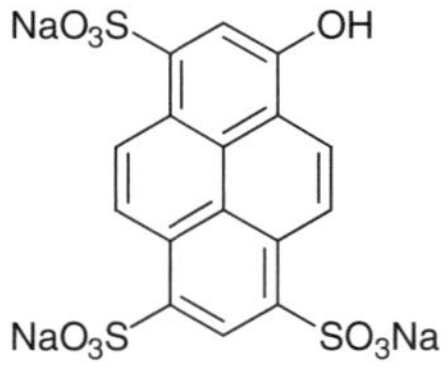

8-hydroxypyrene-1,3,6-trisulfonic acid

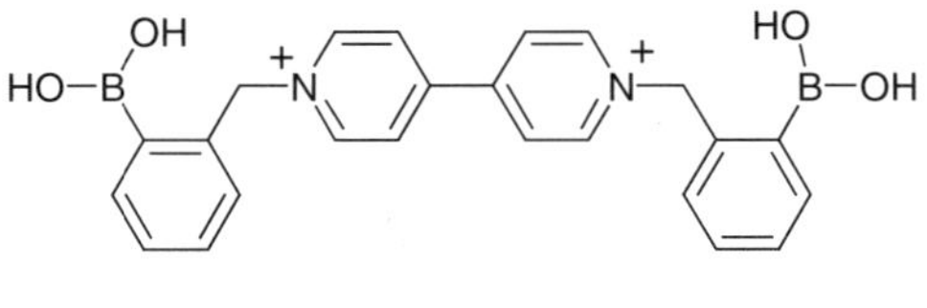

196

197

An interesting multi-component system has also been devised by Singaram, where quenching of a anionic pyranine dye by bisboronic acid viologen units **195**, **196** and **197** is modulated by added saccharide.[324–326] Compound **195** binds well with D-fructose (K_{obs}=2600 M^{-1}) and weakly with D-glucose (K_{obs}=43 M^{-1}) in pH 7.4 phosphate buffer,[325] compound **196** also binds weakly with D-glucose,[326] but more importantly both systems only produces a 4–6% fluorescence recovery. Whereas, compound **197** binds well with both D-fructose (K_{obs}=3300 M^{-1}) and D-glucose (K_{obs}=1800 M^{-1}) in pH 7.4 phosphate buffer. Together with the enhanced selectivity for D-glucose, this system also produces a 45% fluorescence recovery on addition of saccharides.[324]

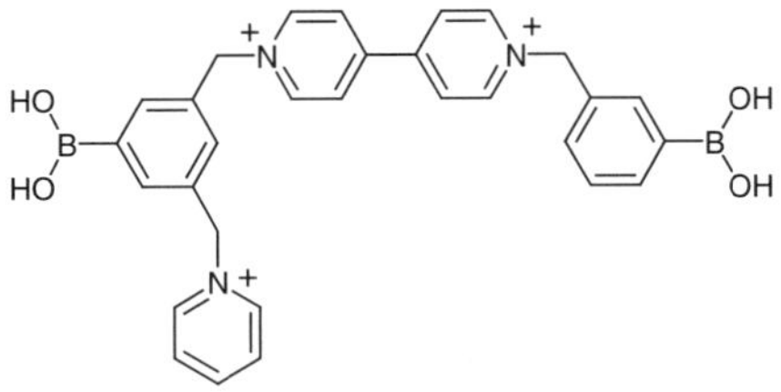

198 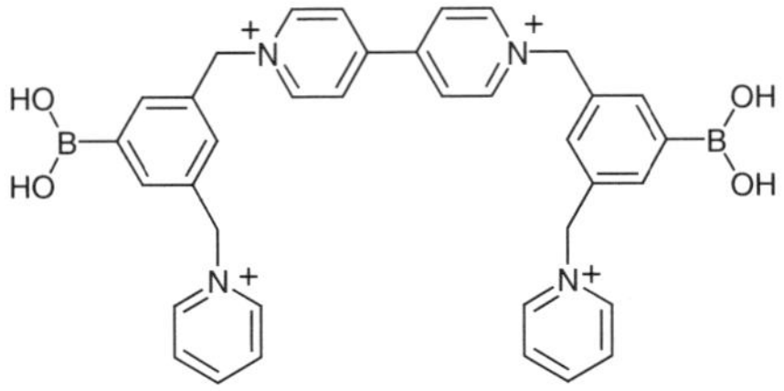   **199**

Through compounds **198** and **199** the effect of increasing the charge of the viologen unit **196** was investigated. The higher the charge on the viologen unit the greater the quenching of the anionic dye, *i.e.*, interaction of the anionic dyes with higher charged species is greater. However, the glucose-sensing ability is reduced since dissociation of the viologen-dye duplex is made more difficult.[327]

The effect of varying the anionic dye component used with viologen **199** was also investigated. The system employed 11 different anionic dye molecules and determined that anionic dyes with higher charge worked better as saccharide sensory systems.[328]

200

Lakowicz has also examined the quenching and recovery of a sulfonated poly(phenylene ethynylene) by a *bis*-boronic acid viologen **200** on addition of saccharides.[329] The system is D-fructose selective and produces up to 70-fold fluorescence enhancement on addition of saccharides.

Wang has recently shown that alizarin red S and phenylboronic acid (PBA) could be used in competitive assays for saccharides.[67,330,331] The system is D-fructose selective, which is the expected selectivity for a monoboronic acid system.[64] This system takes advantage of the known interaction of alizarin red S with boronic acids.[332] The observed stability constants (K_{obs}) for the PBA alizarin red S assay were 160 M^{-1} for D-fructose and 4.6 M^{-1} for D-glucose in water at pH 7.4 (phosphate buffer). Developing this elegant system Hu employed 3-pyridinylboronic acid and pyrocatechol violet in a competitive assay for D-glucose.[333] The observed stability constants (K_{obs}) for assay were 272 M^{-1} for D-glucose in water at pH 7.4 (phosphate buffer). Therefore, this very simple system can be used to detect millimolar D-glucose.

201 (*R*)

Alizarin Red S

James and co-workers have also used alizarin red S in the design of a D-glucose selective-fluorescent assay. The receptor **201** was based on the successful fluorescent PET sensor **143**$_{(n=6)}$.[334] Sensor **201** and alizarin red S show a six-fold enhancement over PBA for D-glucose. Sensor **201** can also be used at a concentration 10 times lower than PBA. The observed stability constants (K_{obs}) for **201** were 140 M^{-1} for D-fructose and 66 M^{-1} for D-glucose in 52.1 wt % methanol/water at pH 8.21 (phosphate buffer).

Alizarin red S has also been used by Basu with a number of commercial monoboronic acids, it was found that 3-methoxycarbonyl-5-nitrophenylboronic acid, was much more efficient than PBA in competitive assays.[335] The observed stability constants (K_{obs}) for 3-methoxycarbonyl-5-nitrophenylboronic acid were 1350 M^{-1} for D-fructose in 2.5–3.8% (v/v) THF/water at pH 7.5 (PBS buffer).

From what has been described above, the fluorescent assay method seems to represent one of the best ways forward in the design of saccharide-selective sensors. However, in a competitive assay all competition must be controlled so that the signal can used to produce an analytical outcome. The presence of previously unrecognised interactions between boronic acids and buffer conjugate bases (phosphate, citrate and imidazole)[72] to create ternary complexes (boronate-X-saccharide) will need to be considered in future assay design.

6.4   Polymer and Surface Bound Sensors

If practically useful sensors are to be developed from the boronic acid sensors described so far, then they will need to be integrated into a device. One way to help achieve this goal is to incorporate the saccharide selective interface into a polymer support.

202

Smith has prepared a grafted polymer containing a ribonucleoside 5′-triphosphate selective sensor. The polymers were prepared using poly(allylamine) (PAA) to which 10% of boronic acid monomer unit **202** was grafted.[336] Also, a library of potential sialic acid receptors was prepared.[252] In this case, the polymers were prepared using PAA to which 2% of the boronic acid monomer unit **202** was grafted. The final polymers also contained various amounts of 4-hydroxybenzoic acid, 4-imidazolacetic acid, octanoic acid and/or succinic anhydride.

Nagasaki and Kimura have used polymers of poly(lysine) with boronic acids appended to the amine residue as saccharide receptors.[337–339] On saccharide complexation, these polymers are converted from neutral sp^2 boron to anionic sp^3 hybridised boron. The anionic polymer thus formed interacts with added cyanine dye. Saccharide binding can then be "read-out" by changes in the absorption and induced circular dichroism (ICD) spectra of the cyanine dye molecule.

203

204

Wang has employed the template approach using monomer **203** to prepare a fluorescent polymer with enhanced selectivity towards D-fructose.[340,341] Appleton has used a similar approach using monomer **204** to prepare a D-glucose selective polymer.[240] The Appleton polymer clearly shows the value of the imprinting technique. Here, the selectivity of the monomer for D-fructose over D-glucose has been reversed in the polymeric form.

205

James and co-workers have developed polymer sensors by grafting a solution-based D-glucose selective receptor $143_{(n=6)}$ to a polymer support.[242] The major difference between the polymer-bound system **205** and solution-based system $143_{(n=6)}$ is the D-glucose selectivity. The D-glucose selectivity drops for polymer **205**, whereas the selectivity with other saccharides is similar to those observed for compound $143_{(n=6)}$. However, the polymeric system still has enhanced D-glucose selectivity (nine times) over the monoboronic acid model compound. The reduced binding of **205** for D-glucose has been attributed to the proximity of receptor to the polymer backbone.

206

Kawanishi has also developed a membrane in which the PET-based D-glucose sensing system **80** has been immobilised to generate the polymer system **206**. The amide group was introduced not only as a linker to the membrane but also to shift the excitation and emission maxima to longer wavelengths. The fluorescent boronic unit has also been attached to the membrane through two linkers (*via* the two amino nitrogens). However, with two linkers the fluorescence response and affinity for D-glucose were reduced. These results indicated that single-chain immobilisation is superior to double-chain immobilisation. It was also confirmed that **206** is capable for the sensing of D-glucose in blood.[342]

R^n = amino acids

207

Anslyn has prepared boronic acid polymers **207** incorporating one of 19 natural amino acids (cysteine excluded) at each of three sites on two different binding arms. Out of this library of 19^3 (6859), unique receptors 29 resin beads were randomly selected and used in an assay capable of differentiating between various proteins.[343]

208

Singaram has made a significant breakthrough in the development of a continuous D-glucose-monitoring system by incorporating his assay system, comprising of a pyranine dye and bisboronic acid viologen units, into a thin film hydrogel **208**.[326,344] The system is able to detect D-glucose in the physiologically important range of 2.5–20 mM and operate reversibly under physiological conditions, *i.e.*, 37°C, 0.1 μM ionic strength and pH 7.4.

Asher has developed an attractive system using a crystalline colloidal array (CCA) incorporated into a polyacrylamide hydrogel and created a polymerised crystalline colloidal array (PCCA). Two systems have been developed, one is a polyacrylamide that has pendant boronic acid groups and works in low ionic strength solutions.[345] The other has pendant polyethylene glycol or 15-crown-5 and boronic acid groups and works in high ionic strength solutions.[346,347] The embedded CCA diffracts visible light, and the PCCA diffraction wavelength reports the hydrogel volume. The PCCA photonic crystal sensing material responds to D-glucose by swelling and redshifting the diffraction as the D-glucose concentration increases.

Wolfbeis has prepared a polyaniline with a near-infrared optical response to saccharides. The film was prepared by co-polymerisation of aniline and 3-aminophenylboronic acid. Addition of saccharides at pH 7.3 led to changes in absorption at 675 nm.[348]

209

Lakowicz has employed boronic acid-modified silver nanoparticles in the detection of polysaccharides such as dextran. When the nanoparticles couple to multiple sites in the polysaccharide, the proximity causes an increase in luminescence due to surface-enhanced fluorescence.[349,350]

Nakashima has prepared a phenylboronic acid terminated redox active self assembled monolayer on a gold electrode as an electrochemical sensor for saccharides. Self-assembled monolayers of **209**, a phenylboronic acid terminated viologen alkyl disulfide, function as a sensitive saccharide sensor in aqueous solution.[351]

Freund has prepared polyaniline boronic acids by the electrochemical polymerisation of 3-aminophenylboronic acid.[352,353] The electrochemical potential of the polymer is sensitive to the change in the pK_a of the polymer as a result of boronic acid–diol complex formation. Fabre has also used polyaniline boronic acids as a conductiomeric sensor for dopamine.[354]

Several other polymers containing 3-aminophenylboronic acid[355,356] and vinylphenylboronic acid[357] groups have been evaluated as electrochemical saccharide sensors.

6.5 Odds and Ends

Boronic acids have been used in the development of surface plasmon resonance (SPR),[358–360] quartz crystal microbalance (QCM) sensors,[358,361,362] holographic sensors.[363] They have also been used in faradaic impedence spectroscopy,[361] ion-sensitive field effect transistors (ISFET),[364] and chemical exchange saturation transfer (CEST) contrast agents for magnetic resonance imaging (MRI).[365,366] Boronic acids have also been used as chiral derivitising agents to determine the enantiomeric purity of chiral amines.[367] Boronic acid nanoparticle–alizarin red S diads have been used in competitive assays for saccharides.[368] Boronic acids have also been used to immobilise glycosylated enzymes to carbon electrodes.[369] The swelling of phenylboronic acid polymers has also been used to control the release of insulin.[370,371]

Other Systems For Saccharide Recognition

If we knew what it was we were doing, it would not be called research, would it?

Albert Einstein, 1879—1955

7.1   Receptors at the Air–Water Interface

210   **211**   **212**

Molecular assemblies of monolayers and their properties have been well established.[372] The unique characteristics of Langmuir–Blodgett (LB) films have drawn particular attention. The pressure of an LB film is sensitive to the activity of its individual constituents. The boronic acids **210** and **211** were prepared by Shinkai in order to demonstrate the effect of saccharide binding at the air–water interface.[109] The reactivity of the boronic acids was tested by solvent extraction methods (solid–solvent, neutral solvent–solvent and basic solvent–solvent). Extractabilities of both compounds were found in the order D-fructose > D-glucose > D-maltose > D-saccharose. Monolayers formed by **210** were found to be unstable based on both unreproducible pressure–area (π–A) isotherms and the crystalline nature of the monolayer. The meta-isomer, **211**, on the other hand, gave very reproducible results. The π–A isotherm of **211** was affected by the introduction of saccharide into the sub-phase at pH 10. The chiral cholesterylboronic acid derivative **212** investigated by Ludwig behaved similarly.[373] Due to its chiral nature it could selectively recognise chiral isomers of fructose. The influence of quaternised amines on the saccharide binding has also been investigated by Ludwig.[374] Quaternised amines facilitate the saccharide detection by the monolayer at neutral pH. Assistance of closely located ammonium cations in the formation of boronate anion is believed to be the source of enhancement.

213: Methyl
214: R = 2-Octyldodecyl
215: R = 4-tert-Butylbenzyl
216: R = -CH$_2$CH=C(CH$_3$)CH$_2$CH$_2$CH=C(CH$_3$)$_2$ (Geranyl)

The cooperative binding of saccharides by the diboronic acid derivatives **213**, **214**, **215** and **216** on monolayers has also been investigated by Ludwig and found to be in agreement with its recognition pattern in homogeneous solutions.[375,376] Molecular recognition in this system has been studied by Dusemund and also seems to be facilitated by closely located ammonium cations.[376]

Recently, Kurihara has prepared electroconductive LB films for sugar recognition. A polymerised electroconductive boronic acid LB film was transferred onto a gold surface, binding with an electroactive manoside (nitrobenzene) was monitored by cyclic voltammetry.[377]

Pietraszkiewicz has investigated the interaction of L-sorbose, D-glucose, D-galactose and D-cellobiose with Langmuir films of boron-containing cavitands derived from calix[4]resorcinarenes.[378] Binding could be detected but, they were unable to make any conclusions regarding the stereochemistry or stoichiometry of the complexes formed.

7.2   Transport and Extraction

The aim of a saccharide extraction is to selectively bind with a saccharide in water and then be able to move the saccharide into a hydrophobic solvent (membrane). Saccharide transport is very similar to extraction; a saccharide must be bound in water then moved into a hydrophobic membrane. Once in the membrane the saccharide must be transported across the membrane and then released into the water on the other side of the membrane. Although the properties that are required for a good molecular extractor and transporter are similar, good transporters must balance extraction with release.

Efficient transport of saccharide–related, water-soluble artificial drugs into individual cells *via* the cell membrane are critical to the future development of drug design and delivery. Many biomimetic systems, which are capable of transporting neutral molecular species, are known, although examples of systems that can transport such species actively are rare.[379]

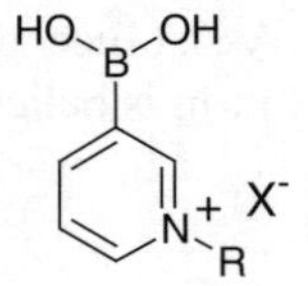

217: R = CH$_3$ X = OMe
218: R = n-C$_{18}$H$_{37}$ X = OTs
219: Cholesterol   X = OMs

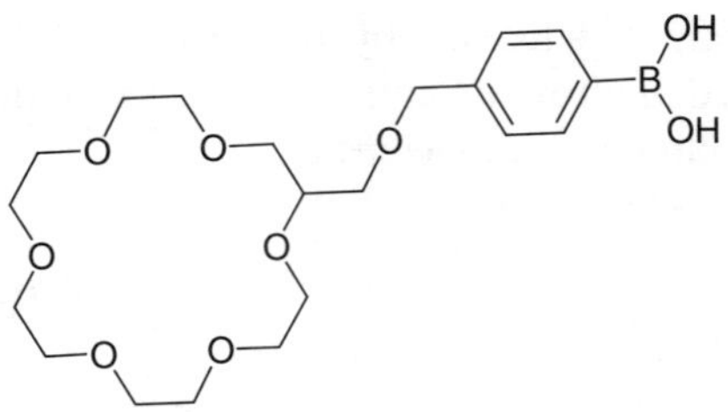

220

Boronic acid and its derivatives have been used as a carrier in the transport of saccharide through membranes.[380] As formation of the anionic boronate is favoured in alkaline pH and disfavoured at lower pH, this provides a means of actively transporting glucose by a pH gradient. In a study of saccharide transport by phenylboronic acid derivatives, phenyl boronate ion has been accompanied by the lipophilic trioctylmethylammonium (TOMA) cation.[380] Not only saccharides but also saccharide-related biologically important molecular species such as uridine have been found to be effectively transported in this way.[381,382] The negatively charged boronate complex accompanies lipophilic cation (trioctylmethylammonium chloride, TOMAC) during the transport. Smith has used the assistance of F-ions in saccharide transport.[383] Reaction of fluoride ions with boronic acid to form phenylfluoroboronate has been proposed as the reason for this observation. This provides a means of active transport of saccharide-related molecular species at neutral pH with a fluoride ion gradient. Interestingly, other halogens do not assist in the transport. The incorporation of a cationic charge into the boronic acid could waive the requirement of an accompanying ammonium cation, such as pyridinium derivatives **217**, **218** and **219**. Czarnik has shown that the strongly acidic pyridinium boronic acid **219** transports saccharide-related species through membranes.[384] The transport of dopamine and related derivatives by the molecular receptor **220** has been reported by Smith.[382] The cooperative binding by two different receptor sites is possible in this case. Transport of amino acid derivatives by boronic acid also has been reported by Czarnik.[385]

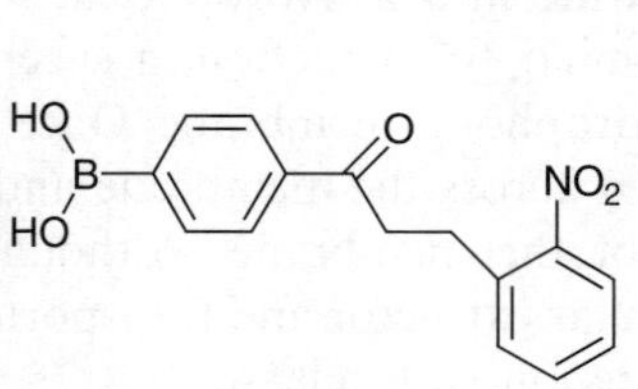

221

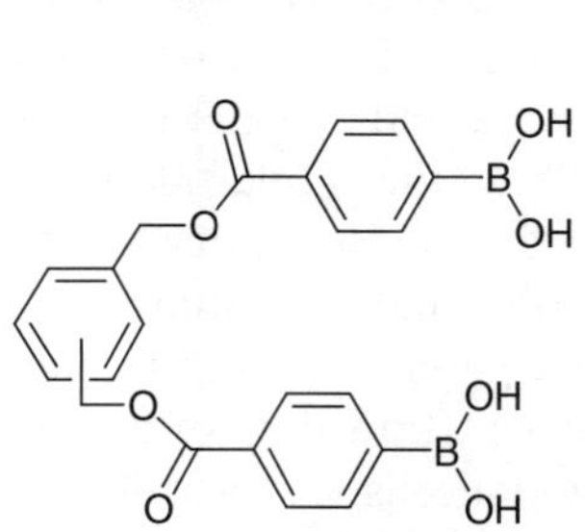

222: *ortho*
223: *meta*
224: *para*

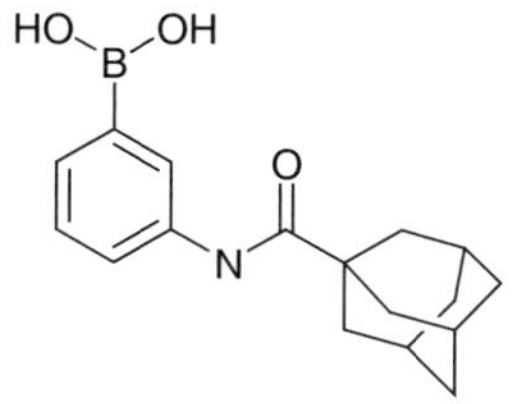

225

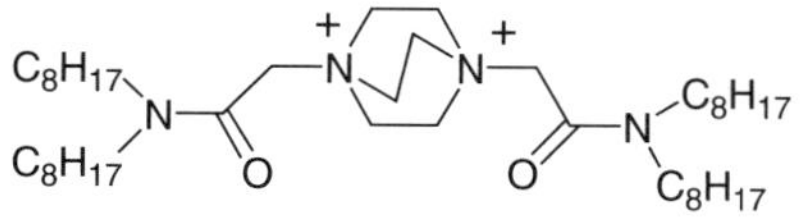

226

Smith has investigated the ability of 21 monoboronic acids to transport saccharides through lipid bilayers.[272] It was found that lipophilic boronic acids are capable of facilitating the transport of monosaccharides through lipid bilayers, but that disaccharides are not transported. The mechanism of transport requires complexation of the saccharide as the tetrahedral boronate. However, the transported species is the neutral conjugate acid. Smith has also investigated selective fructose transport through supported liquid membranes using both mono- and diboronic acids **221**, **222**, **223** and **224**.[271,386] Smith has also used a monoboronic acid **225** in combination with a diammonium cation **226** to facilitate the transport of bororibonucleoside-5′-phosphates.[387]

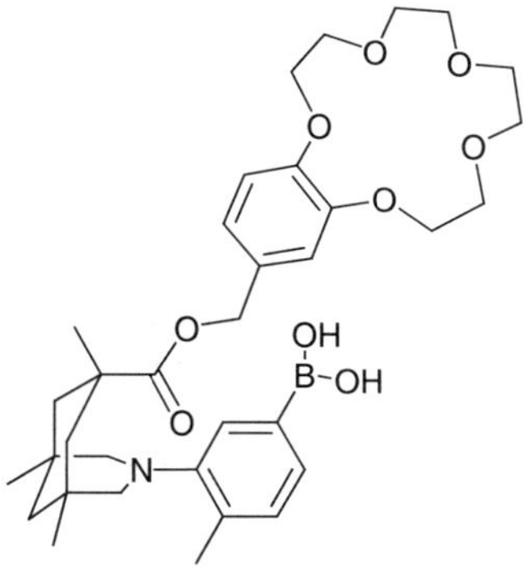

227

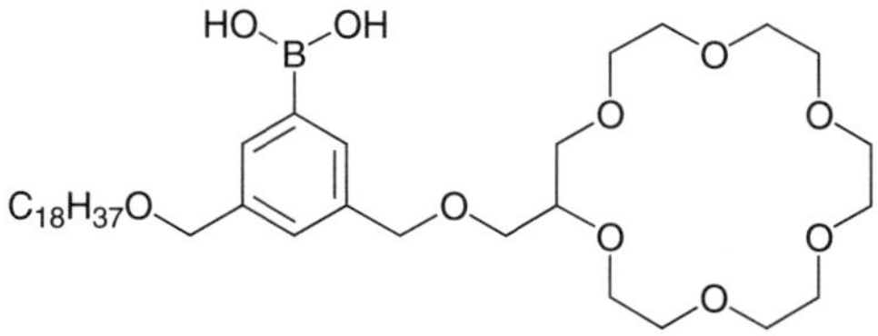

228

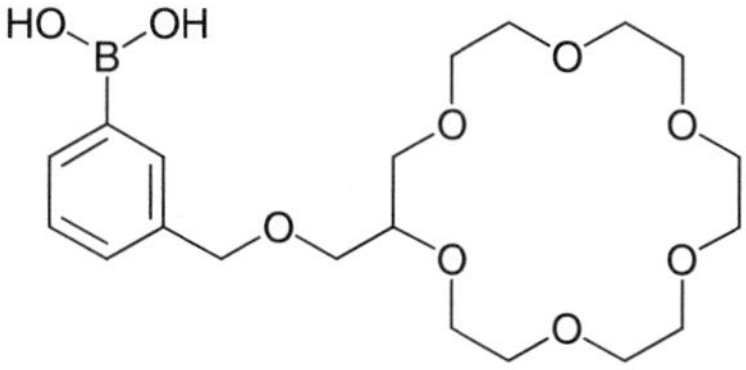

229

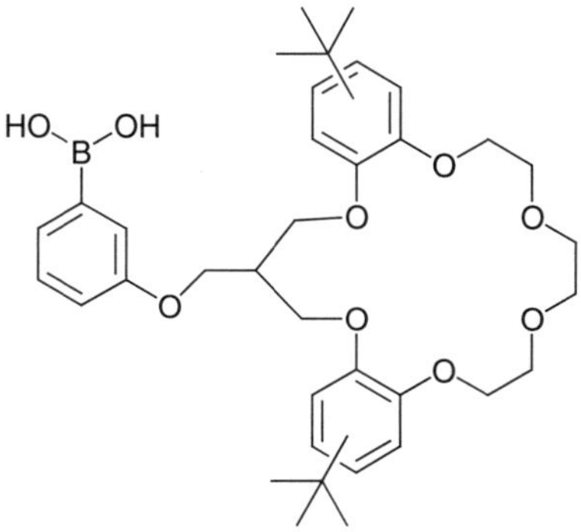

230

Boronic acids with linked ammonium ions have been used by Takeuchi to aid saccharide extraction.[388] Using boronic acids in combination with crown ethers, Smith has developed a sodium-saccharide co-transporter **227**[389] and facilitated catecholamine transporters **228**, **229** and **230**.[390] Smith has used microporous polypropylene impregnated with a boronic acid in 2-nitrophenyl octyl ether as a supported liquid membrane for the separation of fructose from fermentation broths.[391]

231   **232**

Duggan has developed a highly lipophilic boronic acids **231** and **232**, which are able to transport fructose with very high selectivity.[270,392–394] Duggan has achieved unprecedented fructose selectivity by using a rigid five cavitand rim-appended with boronic acid receptor groups.[395]

7.3 CD Receptors

Optical activity stemming from chirality is manifested by practically all natural products such as nucleic acids, sugars, proteins, *etc*. Chiroptical properties have been known since the mid-nineteenth century and are among the most important physical properties for both the food- and drug-related industries. Increasingly strict regulations have applied pressure on industry to produce enantiometrically pure drugs because of the possible devastating deleterious effects of the wrong enantiomer. The induced CD properties of molecular complexes with chiral guest species such as saccharides upon chiral induction are important in providing a method to determine the chirality of the guest. This is significant for non-chromophoric host molecules.

The presence of hydroxy groups with slight reactivity differences creates problems in designing saccharide receptors. However, the high number of hydroxyl groups is also an advantage if these hydroxy groups are used in the recognition/binding process.

7.3.1 Homogeneous Systems

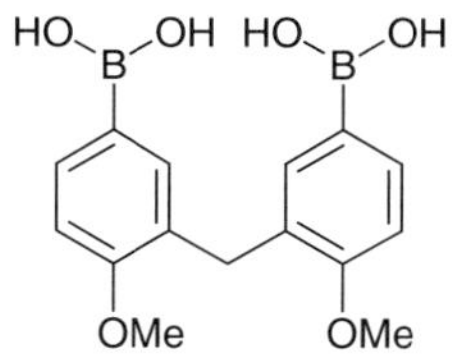

233

When two boronic acid units are arranged in a specific orientation, a saccharide may be bound in a 1:1 fashion by its head and tail. The cooperative binding of two boronic acids creates a rigid cyclic complex. The asymmetric immobilization of two chromophoric benzene rings by ring closure with saccharides can be read out by CD spectroscopy. Diboronic acid **233** prepared by Shiomi forms small cyclic complexes with monosaccharides and some disaccharides in basic aqueous media as monitored by CD spectroscopy.[212,213] (Scheme 44) D- and L-saccharides give positive and negative exciton coupling, respectively, with the highest association constant being 19,000 M^{-1} for glucose. Among the D-saccharides tested, D-galactose was the only saccharide to show a negative exciton coupling in the CD spectrum. Table 10 summarises the CD spectral data and association constants of different saccharides with **233**. This work has shown that the absolute configuration of saccharides can be conveniently predicted from the sign of the CD spectra of **233**. The 1,2-diol is widely

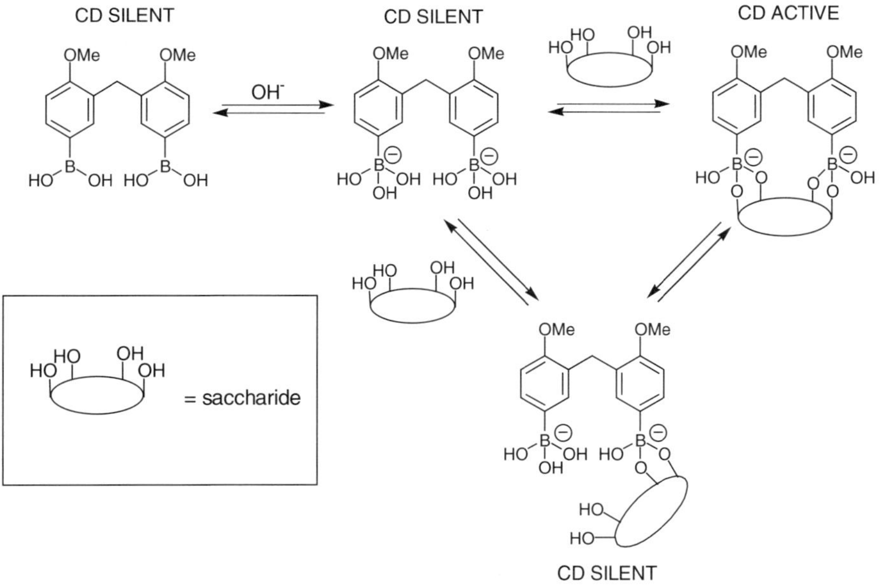

Scheme 44 *Species present in solution responsible for the CD activity observed when* **233** *interacts with saccharides at pH 11.3.*

Table 10   *Absorption and CD maxima of* **233** *monosaccharide complexes[a]*

Saccharide	UV	CD		Stoichiometry	K
	λmax (nm)	λmax (nm)	$[\theta]max^{b}$ (deg cm^2 dmmol^{-1})		(dm^3 mol^{-1})
D-glucose	274	275	−5300	1:1	19000
		205	+231000		
	200	190	−214000		
D-mannose	272	274	−400	1:1	60
		205	+69000		
	200	191	−23000		
D-galactose	273	276	+410	1:1	2200
		205	−22000		
	200	191	+19000		
D-talose	272	275	−3700	1:1	4600
		205	+247000		
	200	190	−196000		
D-fructose	274	$-^{c}$	$-^{c}$	$-^{c}$	$-^{c}$
	200	$-^{c}$	$-^{c}$		

[a] 25°C, [**233**]=1.00 or 2.00 × 10^{-3} mol dm^{-3}, pH 11.3 with 0.10 mol dm^{-3} carbonate buffer.
[b] $[\theta]max$ Values are calculated for 100% complexation values from the 1:1 complex region.
[c] No CD observed.

accepted as the primary binding site of glucose (α-form) with boronic acid.[64] The pyranose structure of the D-glucose diboronic acid **233** complex was suggested using 1H NMR. However, a recent investigation into the structure of this complex by 11B, 13C and 1H NMR analyses suggests that the complexes contain the furanose form of D-glucose.[233]

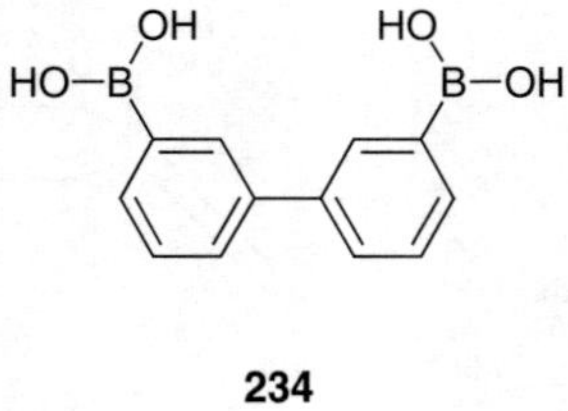

234

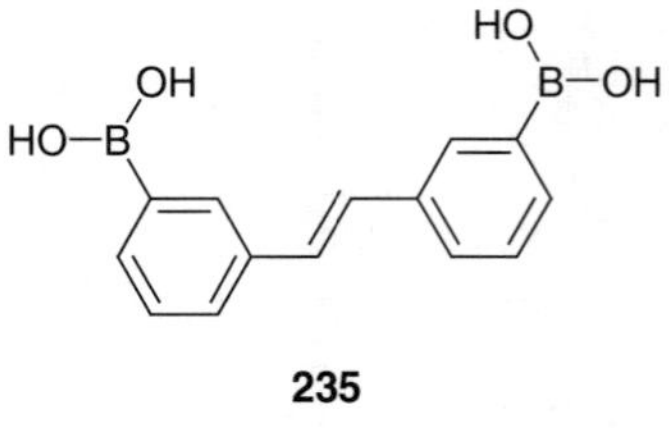

235

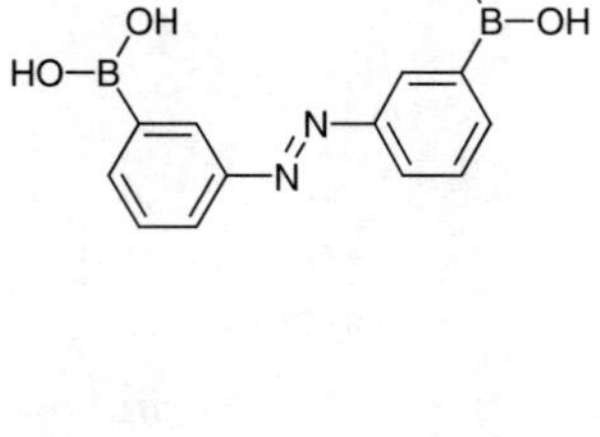

236

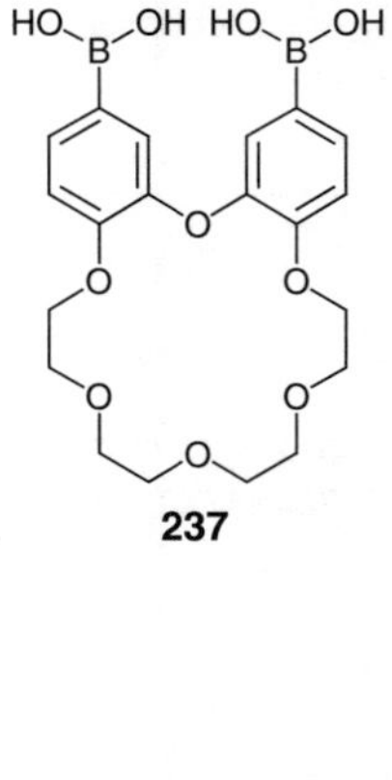

237

Kondo prepared diboronic acid **234**, which has a greater distance between the two boronic acid moieties and was expected to give a strong association with disaccharides.[112,212] However, association constants for disaccharides were much less than those for the complex of monosaccharides and **233**. CPK models revealed that the distance between the two boronic acids in **234** is shorter than the distance between the head and tail hydroxy groups of the disaccharides. Accordingly, the CD activity was much weaker and was inadequate for the estimation of association constants. The 3,3′-stilbenediboronic acid species **235** of Sandanayake similarly gave a CD active species upon disaccharide complexation.[214] Diboronic acid **236**, prepared by James, which has a larger spacer between the two boronic acid moieties, gave better complexes with disaccharides.[211] The extended chromophore facilitates the detection of CD bands and the allosteric switching of selectivity by *cis-trans*-isomerization of **236** has been investigated.

Nature relies on allosteric (or co-operative) interactions in many biological functions. The transport of sugars across a cell membrane into the cell is controlled by such interactions. In an attempt to mimic these functions, a diboronic acid saccharide-binding unit was combined with simple crown ether. The strong cooperative binding of two prompted Deng to prepare allosteric system **237**.[210] The control of the angle between two phenyl rings of **233** by some secondary effect could change the sugar-binding ability of **233**. Having a very similar binding site to **233**, and with greater flexibility in rotation of the phenyl rings, **237** exhibits an even larger association constant for glucose ($31,000\ M^{-1}$) in basic methanolic aqueous solutions (CD measurements). The binding of metal ions to the crown ether was monitored by 1H NMR spectroscopy. Metal ion binding to the crown ether induces a twist in the crown ether resulting in negative allosterism.

238

Nakashima prepared diaza 18-crown-6 diboronic acid **238**.[396] It was shown that saccharides and calcium ions interact competitively for the receptor.

239   **240**

Bipyridine (bpy) diboronic acid **239** developed by Nakashima and its iron (II) (as FeCl2) complex produce CD-active saccharide complexes.[397] The CD activity of **239** was derived from the asymmetric immobilization of the two-pyridine units on saccharide binding. The copper(II) complex gave a CD band in the region of the metal to ligand charge transfer band. Δ or Λ complexes were found depending on the complexed saccharide. For example, the D-maltose complex adopted Λ chirality whereas the D-cellobiose complex adopted Δ chirality.

Mizuno prepared similar complexes with cobalt (II) and bipyridine (bpy) diboronic acid **239**.[398,399] The diboronic acid saccharide complex can be used to control the chirality of the cobalt (II) bipyridine complex. The system is then converted to the substitution-inactive cobalt (III) and the boronic acids removed using silver nitrate to give $[Co^{III}(bpy)_3]^{3+}$.[398,399]

Nuding used saccharides and **240** to transcribe chirality into metal complexes.[400] Yamamoto demonstrated that the helicity of the copper(I) complex of **97** is also controlled by added saccharide.[401,402]

1,2-alternate   cone

241

When large Rb^+ or Cs^+ ions are added to calix[4]arene **241** prepared by Ohseto, it adopts a "1,2-alternate" conformation due to the predominant cation–π interactions.[403] In this conformation, the two boronic acid groups are arranged at a distance suitable for the D-glucose binding. But when a small Li^+ or Na^+ ion is added the calix[4]arene **241** adopts a "cone" conformation due to the predominant metal–oxygen interactions and in the "cone" form D-glucose binding to the diboronic site is suppressed.

242

Imada has developed a novel glucose-6-phosphate selective system **242** based on a boronic acid appended metalloporphyrin derivative.[404,405] The two-point binding of glucose-6-phosphate creates a rigid complex, which gives a strong exciton coupling signal in the CD spectrum. It is believed that the strong binding by the primary binding site of glucose (1,2-diol) to the boronic acid followed by the secondary interaction of the phosphate-metal centre results in the high affinity for glucose-6-phosphate. The replacement of the 1-hydroxy group, which is part of the primary binding site of glucose, by a phosphate unit in glucose-1-phosphate weakens the saccharide–boronic acid interaction and hence the overall strength of the complex. Binding of phosphate to the metal center was revealed by ^{31}P NMR peak shifts. CD exciton coupling peaks were not found for either metal ion free **242** with glucose phosphate or for **242** with glucose itself, indicating the importance of the two-point binding.

243   **244**

Takeuchi explored a very interesting system consisting of a monoboronic acid zinc porphyrin **243** and monoboronic acid pyridine unit **244**.[406] Self-assembly of the porphyrin and pyridine results in a "diboronic acid". This system is particularly interesting since the synthesis of monoboronic acids is much easier than that of diboronic acids. Modification of each constituent monoboronic acid is therefore much easier than that required in the direct synthesis of a diboronic acid. This system will allow for a large number of structurally similar self-assembled diboronic acids to be investigated with the minimum of synthetic effort. Also, using **244** and dicatechol porphrin Sarson has developed a novel porphrin assembly.[407]

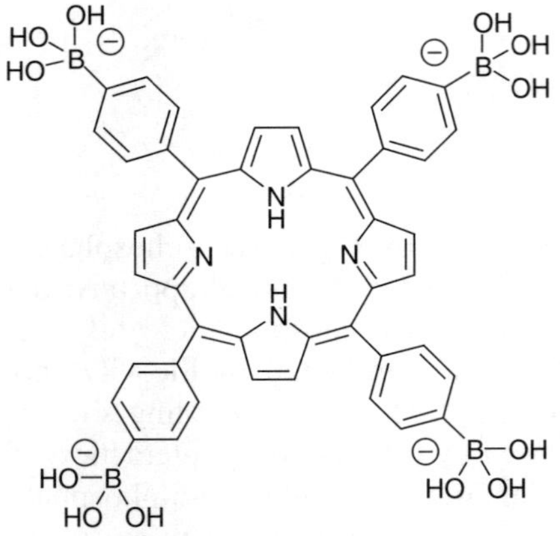

245

Arimori employed a novel dimeric system using anionic porphyrin **245** and cationic porphyrins **193** and **194**.[408] The 1:1 dimer formed between **245** and **193** or **194** showed selective binding with glucose and xylose. Suenaga has also used porphyrin **193** in saccharide controlled intercalation with DNA.[409] With no added saccharide **193** intercalates with DNA but when saccharides are added **193** dissociates from DNA.

Takeuchi has prepared novel porphyrin dimers using monoboronic acid porphyrin and saccharides.[410,411] These systems show saccharide-controllable electron-transfer efficiency.

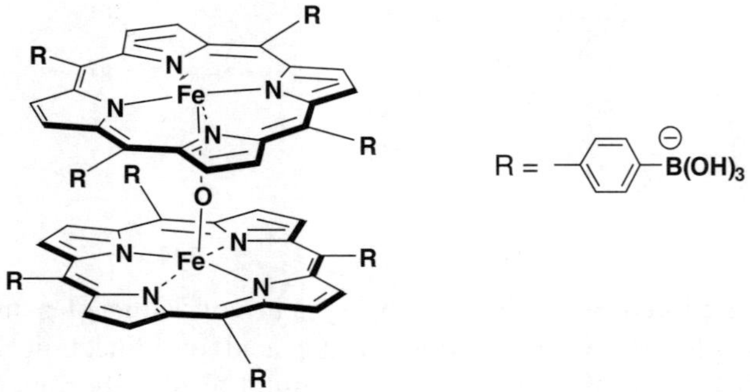

246

The tweezer like μ-oxodimer of iron porphyrin **246** developed by Takeuchi shows very strong and selective binding with D-glucose and D-galactose.[412,413]

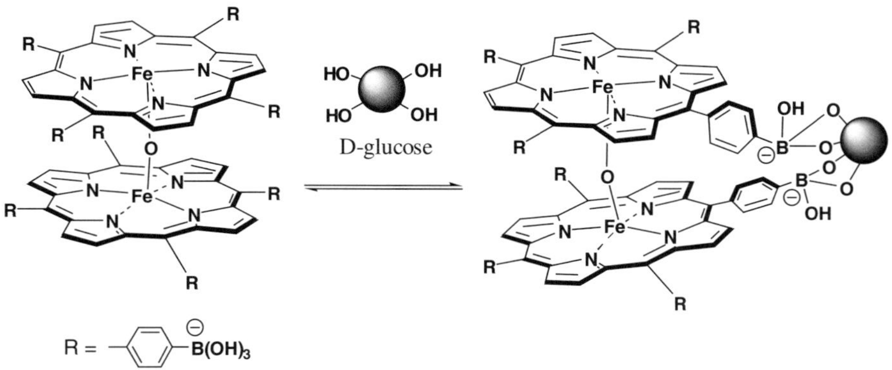

Scheme 45 *Saccharide "tweezer" formed from the μ-oxodimer of iron porphyrin* **246**.

When two boronic acids bind with a saccharide they must get close to each other and the Fe–O–Fe bond angle is tilted to 150° from the regular 180° bond angle. As a result, the distances between the two boronic acids in the remaining three pairs become too long to complex saccharides intramolecularly (Scheme 45). From plots of CD intensity (θ at 380 nm) *vs.* [saccharide], binding constants (K_{obs}) are 1.51×10^5 M^{-1} for D-glucose and 2.43×10^4 M^{-1} for D-galactose.

247

Sugasaki developed the cerium(IV) *bis*(porphyrinate) double decker porphyrin **247** as a scaffold for the recognition of oligosaccharides such as sialyl Lewis X.[247,248] Binding of the first guest species suppresses the rotational freedom of two porphyrin planes and arranges the residual binding sites perfectly for the second guest. The positive homotropic allosterism results in sigmoidal guest response and enhanced affinity and selectivity.

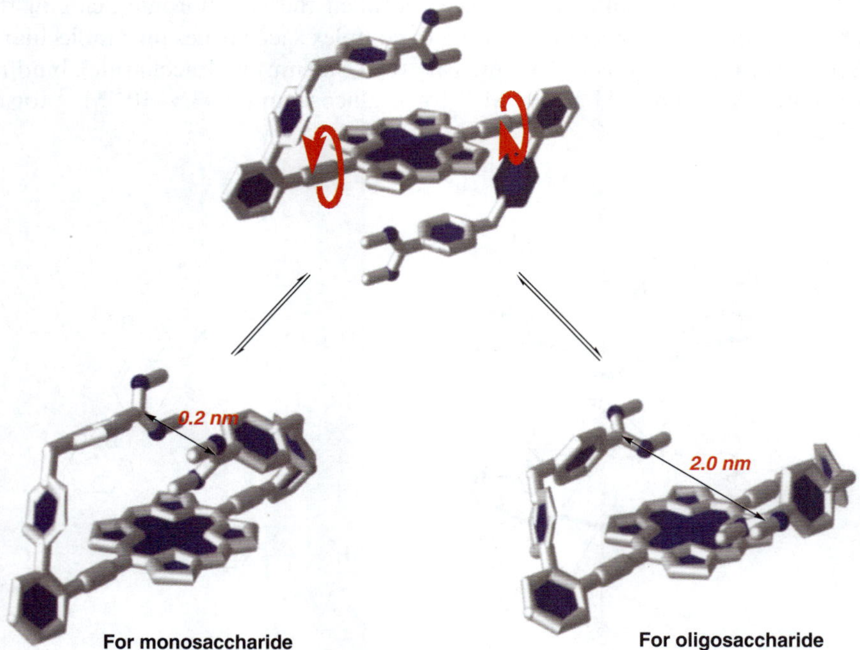

248

Zinc porphyrin **248** developed by Hirata contains two boronic acid arms, which can rotate around their ethynyl axes.[414] This rotation allows the distance between the boronic acid receptors to vary between 0.1 and 2.4 nm (Scheme 46). In particular the zinc complex of **248** produces strong CD signals on binding both monosaccharides and oligosaccharides.

Scheme 46 *Schematic illustration of **248** for mono- and oligosaccharide. The distance between boronic acids can vary for monosaccharide sensing (0.2–0.3 nm) and for oligosaccharide sensing ($\sim$2.0 nm).*

7.3.2   Heterogeneous and Polymeric Systems

As discussed in previous chapters, the saccharide–boronic acid interaction has been employed in the development of sensors. This interaction is also useful as a communication point between saccharides and molecular assemblies. One may regard this connection into two different ways: first, one can sense sugars utilizing some changes in the physical properties of molecular assembly systems and second, one can control the superstructure of molecular assemblies utilizing saccharides as a trigger.

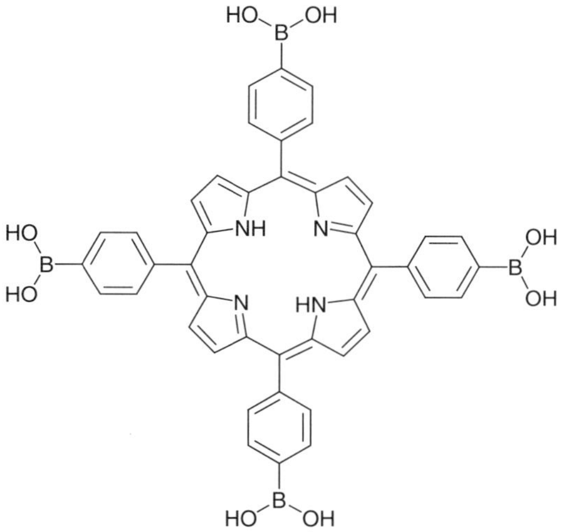

249

Some chiral molecules are known to form chiral helical aggregates.[415] Likewise, chirality can be induced in aggregates of non-chiral molecules by interactions with chiral species.[416,417] Imada found that helical chiral aggregates are found when aggregates of a non-chiral tetraboronic acid porphyrin **249** are treated with monosaccharides.[418] Induced chirality of the aggregate can be monitored by CD spectroscopy. Furthermore, the sign of the exciton coupling of the sugar complexed aggregate can be predicted by the structural orientation of the complex. The chromophoric boronic acid derivative **157** was also found to aggregate in mixed solvents and became CD active when saccharides were added (water:dimethyl formamide).[142,282]

Organogels and hydrogels are formed, mainly, by aggregation of amphiphilic molecules. Thus, a variety of superstructures are constructed in the gel system, reflecting the molecular shape of unit gelator molecules. When boronic acid-appended amphiphiles are dispersed into solution, the superstructure of the resultant aggregates is strongly affected by added sugars.[419–421] For example, the sol-gel phase-transition temperature is changed by the addition of sugars and in some cases, the chirally twisted fibrils appear in the gel phase from achiral amphiphiles. When boronic acid-appended cholesterol is used as a gelator, the gel can discriminate between D- and L-saccharides through a difference in the sol-gel phase-transition temperature.[417]

250

251

Kimura has applied the concept to the structure control of ordered molecular assemblies formed in an aqueous system. Compound **250** bearing a boronic acid group and a chromophoric azobenzene group forms a micelle-like, order-less aggregate in aqueous solution. When saccharides are added a stable membrane is formed.[422] The CD spectrum arising from the azobenzene chromophore suggests that this membrane changes in response to the absolute configuration of added sugars. Arimori developed a similar system using the porphyrin-containing amphiphile **251**.[423,424] In this system, the arrangement of the porphyrin moiety in the molecular assemblies can be finely tuned by added saccharides.

James demonstrated that this idea is effective in a liquid crystal system. It is known that in some cholesteric liquid crystals, the change in the helical pitch is reflected by the colour change. It was found that cholesterylboronic acid **212** complexes of monosaccharides alter the colour of a composite chiral cholesteric liquid crystal membrane.[425] Interestingly, the direction and the magnitude of the colour change is indicative of the absolute configuration of the monosaccharide. This finding makes it possible to predict the absolute configuration of the monosaccharide by a simple ''colour change''.

Kobayashi found that sugar-based amphiphiles can enjoy a fibre-vesicle structural interconversion in the presence of boronic acid-appended poly(L-lysine).[426,427] The research objects of these papers were to design bolaamphiphilic gelators utilizing a sugar family as a source of solvophilic groups and an azobenzene segment as a solvophobic group and to monitor the aggregation mode utilizing the spectroscopic properties of the azobenzene chromophore. The results indicated that the bolaamphiphiles act, although only for specific DMSO-water mixtures, as gelators and form a unique supramolecular helical structure in the gel phase. The UV–Vis and CD spectra showed that the azobenzene segments adopt H-type, face-to-face orientation and the dipole moments are arranged in the right-handed (R) helicity. Since the fibrils as observed by electron microscope possess the right-handed helical structure, one may consider that the microscopic azobenzene–azobenzene orientation is reflected by the macroscopic supramolecular structure. When boronic acid-appended poly(L-lysine) was added, the gel phase of gelator **252** was changed into the sol phase in the macroscopic level and the fibrous aggregate was changed into the vesicular aggregate in the microscopic level. These changes, which are usually induced by a temperature change, are due to the specific boronic acid–saccharide interaction occurring at constant temperature. Interestingly, when D-fructose that shows high affinity with the boronic acid group was added, the sol phase and the vesicular aggregate were changed back to the gel phase and the fibrous aggregate, respectively (Scheme 47). This means that the phase and morphological changes in the sugar-integrated bolaamphiphiles can be controlled reversibly.

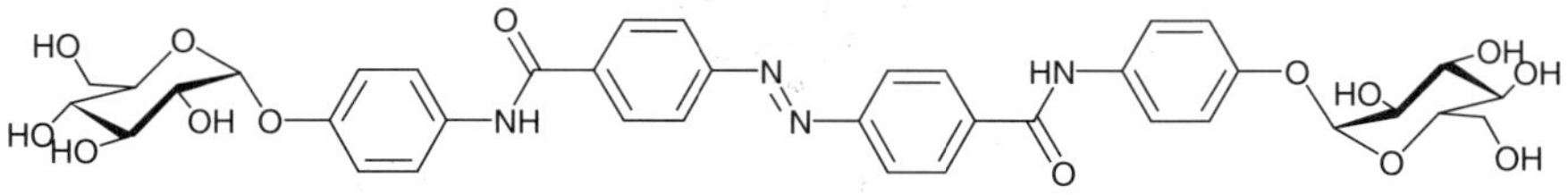

252

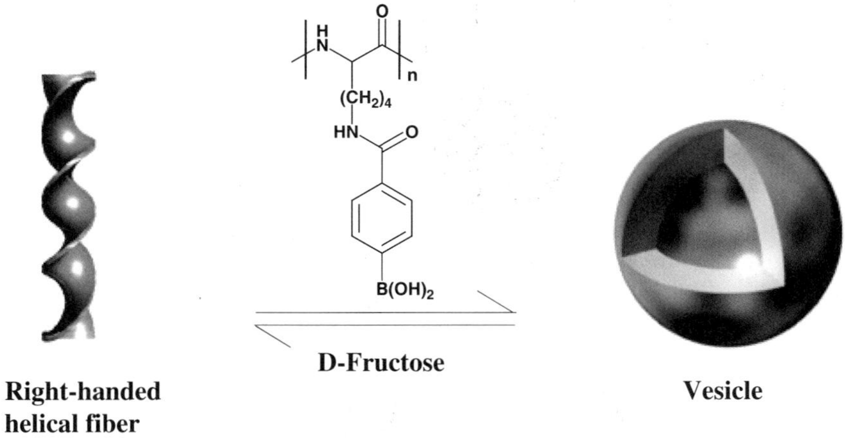

Right-handed helical fiber

D-Fructose

Vesicle

Scheme 47 *Fibre-vesicle interconversion.*

7.4 Molecular Imprinting

Molecular imprinting of saccharides using boronic acids was pioneered by Wulff in 1977.[174,428–430] The approach employed by Wulff produces a 3-D polymer network. Shinkai has also applied the concept of molecular imprinting to homogeneous (0-D), linear polymers (1-D) and at interfaces (2-D). The analysis and preparation of these lower dimensional systems (0-, 1-, 2-D) is much simpler than the 3-D polymeric systems.

Ishi-i has investigated homogeneous imprinting (0-D), [60]fullerene was used as an imprinting base. Two moles of a boronic acid of 1,2-*bis*(bromomethyl)benzene form a complex with one mole of sugar and the resultant 2:1 complex was allowed to react with [60]fullerene in a homogeneous solution (Scheme 48). The

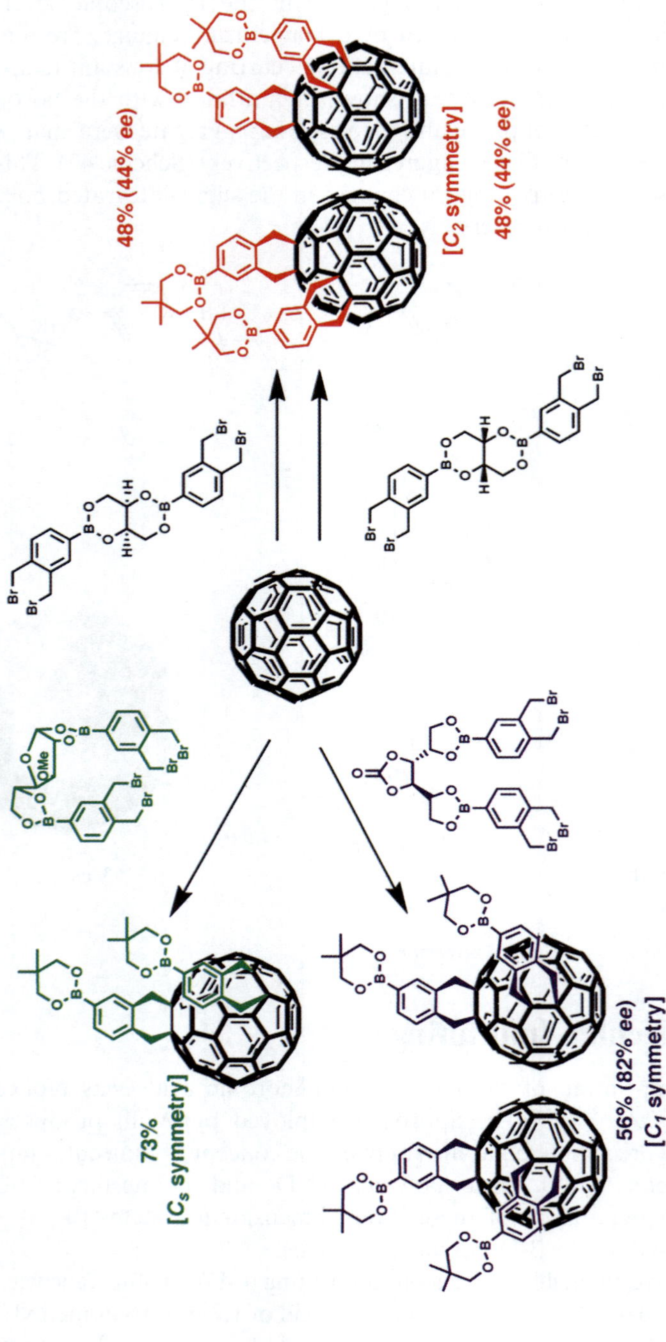

Scheme 48 *Regio- and Chiro-selectiveintroduction of two boronic acid groups into [60]fullerene.*

chiroselectivities attained in this method were very high ($\sim$44–82% e.e.).[431] The diboronic acid/[60]fullerene receptors thus obtained showed a good memory for the templated sugars ($\sim$44–48 % d.e.).[432] The results indicate that the molecular imprinting is possible even in the homogeneous solution, utilizing [60]fullerene containing numerous reactive C=C double bonds.

Kanekiyo explored linear polymer imprinting (1-D), the formation of a 1:1 dimer between anionic polymers and cationic polymers was utilised. The polyanion fragment contains boronic acid units to complex with AMP (Scheme 49). On removal of AMP from the precipitated polyion complex a "cleft" that has the memory for the AMP template is created.[433] It was demonstrated that this cleft shows high affinity with AMP and the precipitate (gel) displays reversible swelling–shrinking in response to the binding of AMP. When this gel is deposited on a QCM resonator, it responds to small changes in the concentration of AMP.[434]

Friggeri investigated imprinting at 2-D interfaces using a QCM resonator coated with Au (Scheme 50).[435] The conformational transitions of poly (L-lysine) are related to very subtle changes in secondary forces such as hydrogen-bonding interactions, electrostatic effects and hydrophobicity.

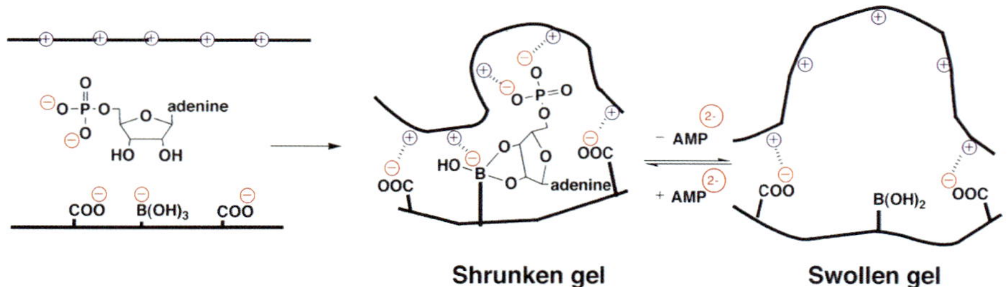

Scheme 49 *AMP-imprinted polyion complex.*

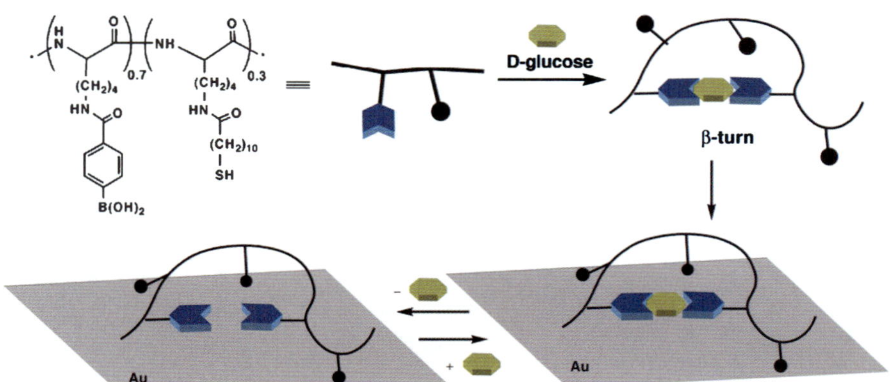

Scheme 50 *Memory creation by molecular imprinting on the Au surface.*

Previously it was discovered that the addition of monosaccharides to boronic acid-appended poly(L-lysine) affects the α-helix content of the polypeptide chain and the pH at which the maximum α-helix content is observed is specific to the added monosaccharide.[339,436] To take advantage of such sugar-induced conformational changes, Friggeri synthesised a poly(L-lysine) derivative appending 70 mol% of the $B(OH)_2$ group and 30 mol% of the SH group. After induction of the conformational change by the sugar–boronic acid interaction, the polymeric complex was extended on Au. The sugar-induced conformation is then immobilised on the Au surface, which keeps a memory for the templated sugar (Scheme 50). Evidence for the imprinting was obtained using an Au-coated QCM resonator.[435] In particular, when D-glucose is used as a template sugar, the polypeptide backbone tends to adopt a unique β-turn structure. Therefore a D-glucose-selective QCM resonator was created by a simple imprinting method.[435]

More recently, it was shown that molecular imprinting can be achieved utilising the re-precipitation process of functionalised "soluble" polymers[437–439] and the gel formation process of low molecular weight gelators.[440] These new trends suggest the generality of the molecular imprinting concept: that is, the formation of 3-D cross-linked resins is not the only way of employing molecular imprinting.

Conclusion

One never notices what has been done; one can only see what remains to be done

Marie Curie, 1867–1934

The development of coherent strategies for the selective binding of saccharides, by rationally designed synthetic receptors, remains one of chemistry's most sought after goals. The research conducted to this end is driven by a fundamental inquisitiveness and need to monitor saccharides of industrial, environmental and biological significance.

This book describes the recent efforts to develop practically useful receptors and sensors for saccharides using boronic acids. Since the first publication on glucose-selective diboronic acid-based PET sensors a decade ago, these systems have proved their worth and have come to find application at the cutting edge of medical care.

Some general observations concerning boronic acid-based saccharide receptors were summarised at the end of Chapter 4. We believe that those observations will help other research groups design new boronic-based receptors for saccharides. Therefore, we will recap some of the more salient points here.

On complexation of a saccharide to a boronic acid there is a contraction of the O–B–O bond angle with a concomitant increase in the acidity of the boron species. Boronic acid–diol complex formation is heavily pH dependent. Rate and stability constants increase by around four and five orders of magnitude, respectively, at pHs above the pK_a of the boronic acid.

The introduction of a carefully located tertiary amine proximal to the boron centre of a fluorescent sensor permits the sensor to function at lower pH and introduces an "off–on" optical response to the system *via* photoinduced electron transfer (PET). The tertiary amine–boronic acid (N–B) interaction in a boronic acid-based PET sensor has strength in the range of 15–25 kJ mol^{-1}. In an aprotic solvent, the N–B dative bond is usually present. However, in a protic media, solvent insertion of the N–B occurs to afford a hydrogen-bonded zwitterionic species.

The use of two boronic acid receptors within a binding site introduces saccharide selectivity, permitting saccharides such as D-glucose to be specifically targeted. Computational data and observed experimental results indicate that strong binding between boronic acids and saccharides occurs preferentially with saccharides that have an available anomeric hydroxyl pair, which has the capacity to conform to a *syn*-periplanar alignment. In the vast majority of cases, this requires formation of the furanose form of the saccharide.

These observations clearly show that many challenges of saccharide recognition can be conquered by using boronic acids. Our hope in preparing this book was that the promise offered by boronic acid receptors would stimulate other groups to develop even better receptors and sensors for saccharides.

You can know the name of a bird in all the languages of the world, but when you're finished, you'll know absolutely nothing whatever about the bird . . . . So let's look at the bird and see what it's doing – that's what counts. I learned very early the difference between knowing the name of something and knowing something.

Richard Feynman (1918–1988)

References

If I have seen further than others, it is by standing upon the shoulders of giants

Isaac Newton, 1642–1727

1. G.R. Desiraju, *Nature*, 2001, **412**, 397–400.
2. B.G. Malmström, Nobel Lectures in Chemistry (1991–1995), Nobel Lectures, World Scientific, 1997.
3. A.P. de Silva, H.Q.N. Gunaratne, T. Gunnlaugsson, A.J.M. Huxley, C.P. McCoy, J.T. Rademacher and T.E. Rice, *Chem. Rev.*, 1997, **97**, 1515–1566.
4. A.W. Czarnik, Fluorescent Chemosensors for Ion and Molecule Recognition, ACS Symposium Series 538, American Chemical Society Books, 1993.
5. T.D. James, P. Linnane and S. Shinkai, *Chem. Commun.*, 1996, 281–288.
6. T.D. James, K.R.A.S. Sandanayake and S. Shinkai, *Angew. Chem., Int. Ed. Engl.*, 1996, **35**, 1911–1922.
7. M.D. Phillips and T.D. James, *J. Fluorescence*, 2004, **14**, 549–559.
8. T.D. James and S. Shinkai, *Top. Curr. Chem.*, 2002, **218**, 159–200.
9. A.P. Davis and T.D. James, *Carbohydrate receptors*, in *Functional Synthetic Receptors*, T. Schrader and A.D. Hamilton (eds), Wiley–VCH, 2005, 45–110.
10. T.D. James, *Boronic acid based receptors and sensors for saccharides*, in *Boronic Acids in Organic Synthesis and Chemical Biology*, D.G. Hall (ed), Wiley–VCH, 2005, 441–480.
11. S. Striegler, *Curr. Org. Chem.*, 2003, **7**, 81–102.
12. W. Wang, X. Gao and B. Wang, *Curr. Org. Chem.*, 2002, **6**, 1285–1317.
13. H. Cao and M.D. Heagy, *J. Fluorescence*, 2004, **14**, 569–584.
14. H. Fang, G. Kaur and B. Wang, *J. Fluorescence*, 2004, **14**, 481–489.
15. M. Granda-Valdes, R. Badia, G. Pina-Luis and M.E. Diaz-Garcia, *Quim. Anal.*, 2000, **19**, 38–53.
16. E.A. Moschou, B.V. Sharma, S.K. Deo and S. Daunert, *J. Fluorescence*, 2004, **14**, 535–547.
17. S. Shinkai and M. Takeuchi, *Biosensors and Bioelectronics*, 2004, **20**, 1250–1259.
18. S. Shinkai and M. Takeuchi, *Bull. Chem. Soc. Jpn.*, 2005, **78**, 40–51.

19. A.P. Davis and R.S. Wareham, *Angew. Chem., Int. Ed.*, 1999, **38**, 2978–2996.
20. E. Klein, M.P. Crump and A.P. Davis, *Angew. Chem., Int. Ed.*, 2004, **44**, 298–302.
21. R. Jelinek and S. Kolusheva, *Chem. Rev.*, 2004, **104**, 5987–6015.
22. P.M. Collins and R.J. Ferrier, *Monosaccharides their Chemistry and their Roles in Natural Products*, Wiley, 1995.
23. R.H. Garrett and C.M. Grisham, *Biochemistry*, 2nd edn, Saunders College Publishing, London, 1999.
24. R.A. Dwek and T.D. Butters, Chem. Rev., 2002, 102, 283–284 (and succeeding articles).
25. T. Yamamoto, Y. Seino, H. Fukumoto, G. Koh, H. Yano, N. Inagaki, Y. Yamada, K. Inoue, T. Manabe and H. Imura, *Biochem. Biophys. Res. Commun.*, 1990, **170**, 223–230.
26. P. Baxter, J. Goldhill, P.T. Hardcastle and C.J. Taylor, *Gut*, 1990, **31**, 817–820.
27. S. de Marchi, E. Cecchin, A. Basil, G. Proto, W. Donadon, A. Jengo, D. Schinella, A. Jus, D. Villalta, P. De Paoli, G. Santini and F. Tesio, *J. Nephrol.*, 1984, **4**, 280–286.
28. L.J. Elsas and L.E. Rosenberg, *J. Clinic. Invest.*, 1969, **48**, 1845–1854.
29. S. Wild, G. Roglic, A. Green, R. Sicree and H. King, *Diabetes Care*, 2004, **27**, 1047–1053.
30. S.P. Marso (ed), The Handbook of Diabetes Mellitus and Cardiovascular Disease, Remidica Publishing, 2003.
31. T. Barnett, *The Insulin Treatment of Diabetes A Practical Guide*, EMAP Healthcare, 1998.
32. The British Diabetic Association Cohort Study, Diabet. Med., 1999, 16, 459–465.
33. I.M. Stratton, E.M. Kohner, S.J. Aldington, R.C. Turner, R.R. Holman, S.E. Manley and D.R. Matthews, *Diabetologia*, 2001, **44**, 156–163.
34. J.S. Cameron and S. Challah, *Lancet*, 1986, **2**, 962–966.
35. D.S. Bell, *Diabetes Care*, 1994, **17**, 213–219.
36. D.E. Bild, J.V. Selby, P. Sinnock, W.S. Browner, P. Braveman and J.A. Showstack, *Diabetes Care*, 1989, **12**, 24–31.
37. Department of Health: London, National Service Framework for Diabetes: One Year On, 2004.
38. W. Gatling, R. Hill and M. Kirby, *Shared Care for Diabetes*, Isis Medical Media, 1997.
39. Collaborative Atorvastatin Diabetes Study, Lancet, 2004, 364, 685–696.
40. The Epidemiology of Diabetes Interventions and Complications Study, J. Am. Med. Assoc., 2003, 290, 2159–2167.
41. The Diabetes Control and Complications Trial Research Group, N. Engl. J. Med., 1993, 329, 977–986.
42. P.M. Collins, *Carbohydrates*, Chapman & Hall, 1987.
43. S.J. Angyal, *Adv. Carbohydr. Chem. Biochem.*, 1991, **49**, 19–35.

44. N.K. Vyas, M.N. Vyas and F.A. Quiocho, *Science*, 1988, **242**, 1290–1295.

45. D.C. Harris, *Quantitative Chemical Analysis*, Freeman, 2003.

46. R.W. Cattrall, *Chemical Sensors*, Oxford Chemistry Primers, Oxford University Press, 1997.

47. A.P.F. Turner, B.N. Chen and S.A. Piletsky, *Clin. Chem.*, 1999, **45**, 1596–1601.

48. Y. Aoyama, Y. Tanaka, H. Toi and H. Ogoshi, *J. Am. Chem. Soc.*, 1988, **110**, 634–635.

49. A.P. Davis and R.S. Wareham, *Angew. Chem., Int. Ed.*, 1998, **37**, 2270–2273.

50. G. Das and A.D. Hamilton, *Tetrahedron Lett.*, 1997, **38**, 3675–3678.

51. S. Anderson, U. Neidlein, V. Gramlich and F. Diederich, *Angew. Chem. – Int. Edit. Engl.*, 1995, **34**, 1596–1600.

52. U. Neidlein and F. Diederich, *Chem. Commun.*, 1996, 1493–1494.

53. A.S. Droz, U. Neidlein, S. Anderson, P. Seiler and F. Diederich, *Helv. Chem. Acta*, 2001, **84**, 2243–2289.

54. E. Frankland and B.F. Duppa, *Justus Liebigs Ann. Chem.*, 1860, **115**, 319–322.

55. A. Michaelis and P. Becker, *Ber. Dtsch. Chem. Ges.*, 1880, **13**, 58–61.

56. A. Michaelis and P. Becker, *Ber. Dtsch. Chem. Ges.*, 1882, **15**, 180–185.

57. E. Khotinsky and M. Melamed, *Ber. Dtsch. Chem. Ges.*, 1909, **42**, 3090–3096.

58. J. Böeseken, *Adv. Carbohydr. Chem.*, 1949, **4**, 189–210.

59. J. Böeseken, *Berichte*, 1913, **46**, 2612–2628.

60. H.G. Kuivila, A.H. Keough and E.J. Soboczenski, *J. Org. Chem.*, 1954, **19**, 780–783.

61. M.I. Wolfrom and J. Solms, *J. Org. Chem.*, 1956, **21**, 815–816.

62. M.F. Lappert, *Chem. Rev.*, 1956, **56**, 959–1064.

63. K. Torssel, *Arkiv. Kemi.*, 1957, **10**, 473.

64. J.P. Lorand and J.O. Edwards, *J. Org. Chem.*, 1959, **24**, 769–774.

65. J.H. Hartley, M.D. Phillips and T.D. James, *New J. Chem.*, 2002, **26**, 1228–1237.

66. S. Soundararajan, M. Badawi, C.M. Kohlrust and J.H. Hageman, *Anal. Biochem.*, 1989, **178**, 125–134.

67. G. Springsteen and B. Wang, *Tetrahedron*, 2002, **58**, 5291–5300.

68. A. Yuchi, A. Tatebe, S. Kani and T.D. James, *Bull. Chem. Soc. Jpn.*, 2001, **74**, 509–510.

69. J. Juillard and N. Gueguen, *Comp. Rend. Acad. Sci. C*, 1967, **264**, 259–261.

70. S. Friedman, B. Pace and R. Pizer, *J. Am. Chem. Soc.*, 1974, **96**, 5381–5384.

71. J.O. Edwards and R.J. Sederstrom, *J. Phys. Chem.*, 1961, **65**, 862–862.

72. L.I. Bosch, T.M. Fyles and T.D. James, *Tetrahedron*, 2004, **60**, 11175–11190.

73. A.E. Martell and R.M. Smith, *Critical Stability Constants*, Plenum Press, New York, 1976.
74. N. DiCesare and J.R. Lakowicz, *Anal. Biochem.*, 2001, **294**, 154–160.
75. S. Friedman and R. Pizer, *J. Am. Chem. Soc.*, 1975, **97**, 6059–6062.
76. R. Pizer and R. Selzer, *Inorg. Chem.*, 1983, **23**, 3023.
77. K. Kustin and R. Pizer, *J. Am. Chem. Soc.*, 1969, **91**, 317–322.
78. L. Babcock and R. Pizer, *Inorg. Chem.*, 1980, **19**, 56–61.
79. R. Pizer and C. Tihal, *Inorg. Chem.*, 1992, **31**, 3243–3247.
80. R.J. Ferrier, *Adv. Carbohydr. Chem. Biochem.*, 1978, **35**, 31–80.
81. S.J. Rettig and J. Trotter, *Can. J. Chem.*, 1977, **55**, 3071–3075.
82. P. Rodriguez-Cuamatzi, G. Vargas-Diaz, T. Maris, J.D. Wuest and H. Hoepfl, *Acta Cryst.*, 2004, **E60**, o1316–o1318.
83. J.H. Fournier, T. Maris, J.D. Wuest, W.Z. Guo and E. Galoppini, *J. Am. Chem. Soc.*, 2003, **125**, 1002–1006.
84. P. Rodriguez-Cuamatzi, G. Vargas-Diaz and H. Hopfl, *Angew. Chem., Int. Ed.*, 2004, **43**, 3041–3044.
85. R.R. Shuvalov and P.C. Burns, *Acta Cryst.*, 2003, **C59**, i47–i49.
86. S.P. Draffin, P.J. Duggan and G.D. Fallon, *Acta Cryst.*, 2004, **E60**, o1520–o1522.
87. C.C. Freyhardt, M. Wiebcke and J. Felsche, *Acta Cryst.*, 2000, **C56**, 276–278.
88. K.L. Bhat, S. Hayik and C.W. Bock, *J. Mol. Struct.*, 2003, **638**, 107–117.
89. K.L. Bhat, S. Hayik, J.N. Corvo, D.M. Marycz and C.W. Bock, *J. Mol. Struct.*, 2004, **673**, 145–154.
90. S. Hayik, K.L. Bhat and C.W. Bock, *Struct. Chem.*, 2004, **15**, 133–147.
91. R.D. Pizer and C.A. Tihal, *Polyhedron*, 1996, **15**, 3411–3416.
92. J. Clayden, N. Greeves, S. Warren and P. Wothers, *Organic Chemistry*, Oxford University Press, 2001.
93. G. Lorber and R. Pizer, *Inorg. Chem.*, 1976, **15**, 978–980.
94. P.J. Duggan and E.M. Tyndall, *J. Chem. Soc., Perkin Trans. 1*, 2002, 1325–1339.
95. G.E.K. Branch, D.L. Yabroff and B. Bettman, *J. Am. Chem. Soc.*, 1934, **56**, 937–941.
96. K. Ishihara, Y. Mouri, S. Funahashi and M. Tanaka, *Inorg. Chem.*, 1991, **30**, 2356–2360.
97. C.Y. Shao, S. Matsuoka, Y. Miyazaki, K. Yoshimura, T.M. Suzuki and D.A.P. Tanaka, *J. Chem. Soc., Dalton Trans.*, 2000, 3136–3142.
98. S. Kagawa, K.-I. Sugimoto and S. Funahashi, *Inorg. Chim. Acta*, 1995, **231**, 115–119.
99. H. Ito, Y. Kono, A. Machida, Y. Mitsumoto, K. Omori, N. Nakamura, Y. Kondo and K. Ishihara, *Inorg. Chim. Acta*, 2003, **344**, 28–36.
100. R. Pizer and P.J. Ricatto, *Inorg. Chem.*, 1994, **33**, 2402–2406.
101. S. Toyota, T. Futawaka, H. Ikeda and M. Oki, *J. Chem. Soc., Chem. Commun.*, 1995, **24**, 2499–2500.
102. J. Grotewold, E.A. Lissi and A.E. Villa, *J. Chem. Soc. A*, 1966, 1034–1037.

103. J. Grotewold, E.A. Lissi and A.E. Villa, *J. Chem. Soc. A*, 1966, 1038–1041.
104. M.M. Kreevoy and J.E.C. Hutchins, *J. Am. Chem. Soc.*, 1972, **94**, 6371–6376.
105. I.M. Pepperberg, T.A. Halgren and W.N. Lipscomb, *J. Am. Chem. Soc.*, 1976, **98**, 3442–3451.
106. D.Y. Lee and J.C. Martin, *J. Am. Chem. Soc.*, 1984, **106**, 5745–5746.
107. M. Yamashita, Y. Yamamoto, K. Akiba and S. Nagase, *Angew. Chem., Int. Ed.*, 2000, **39**, 4055–4058.
108. W.H. Zachariasen, *Acta Cryst.*, 1963, **16**, 385–389.
109. S. Shinkai, K. Tsukagoshi, Y. Ishikawa and T. Kunitake, *J. Chem. Soc., Chem. Commun.*, 1991, 1039–1041.
110. A. Finch, P.J. Gardner, P.M. McNamara and G.R. Wellum, *J. Chem. Soc. A*, 1970, 3339–3345.
111. K. Tsukagoshi and S. Shinkai, *J. Org. Chem.*, 1991, **56**, 4089–4091.
112. K. Kondo, Y. Shiomi, M. Saisho, T. Harada and S. Shinkai, *Tetrahedron*, 1992, **48**, 8239–8252.
113. R.A. Bissell, A.P. Desilva, H.Q.N. Gunaratne, P.L.M. Lynch, G.E.M. Maguire and K. Sandanayake, *Chem. Soc. Rev.*, 1992, **21**, 187–195.
114. M. Böhmer and J. Enderlein, *Chem. Phys. Chem.*, 2003, **4**, 793–808.
115. W.E. Moerner and D.P. Fromm, *Rev. Sci. Instrum.*, 2003, **74**, 3597–3619.
116. W.P. Ambrose, P.M. Goodwin, J.H. Jett, A. Van Orden, J.H. Werner and R.A. Keller, *Chem. Rev.*, 1999, **99**, 2929–2956.
117. M. Sauer, *Angew. Chem., Int. Ed.*, 2003, **42**, 1790–1793.
118. J.R. Epstein and D.R. Walt, *Chem. Soc. Rev.*, 2003, **32**, 203–214.
119. M. Fehr, S. Lalonde, D.W. Ehrhardt and W.B. Frommer, *J. Fluorescence*, 2004, **14**, 603–609.
120. A.P. de Silva, H.Q.N. Gunaratne, T. Gunnlaugsson and M. Nieuwenhuizen, *Chem. Commun.*, 1996, 1967–1968.
121. H.R. He, M.A. Mortellaro, M.J.P. Leiner, S.T. Young, R.J. Fraatz and J.K. Tusa, *Anal. Chem.*, 2003, **75**, 549–555.
122. A.J. Tudos, G.A.J. Besselink and R.B.M. Schasfoort, *Lab. Chip*, 2001, **1**, 83–95.
123. H. Schlebusch, I. Paffenholz, R. Zerback and R. Leinberger, *Clin. Chim. Acta*, 2001, **307**, 107–112.
124. R. Badugu, J.R. Lakowicz and C.D. Geddes, *Biorg. Med. Chem.*, 2004, **13**, 113–119.
125. J. Yoon and A.W. Czarnik, *J. Am. Chem. Soc.*, 1992, **114**, 5874–5875.
126. T.D. James, K. Sandanayake and S. Shinkai, *Angew. Chem., Int. Ed. Engl.*, 1994, **33**, 2207–2209.
127. T.D. James, K.R.A.S. Sandanayake and S. Shinkai, *Nature*, 1995, **374**, 345–347.
128. J.R. Lakowicz, *Principles of Fluorescence Spectroscopy*, Plenum, New York, 1999.
129. M. Kasha, *Discussions of the Faraday Society*, 1950, **9**, 14–19.
130. E. Lippert, W. Lueder, F. Moll, W. Naegele, H. Boos, H. Prigge and I. Seibold-Blankenstein, *Angew. Chem.*, 1961, **73**, 695–706.

131. E. Lippert, W. Lüder and H. Boos, *Advances in Molecular Spectroscopy*, A. Mangini (ed), Pergamon Press, Oxford, 1962.

132. W. Rettig, *Angew. Chem., Int. Ed. Engl.*, 1986, **25**, 971–988.

133. Z.R. Grabowski, K. Rotkiewicz and W. Rettig, *Chem. Rev.*, 2003, **103**, 3899–4031.

134. K. Rotkiewicz, K.H. Grellmann and Z.R. Grabowski, *Chem. Phys. Lett.*, 1973, **19**, 315–318.

135. K. Rotkiewicz, K.H. Grellmann and Z.R. Grabowski, Chem. Phys. Lett., 1973, 21, 212 (Errata).

136. A. Siemiarczuk, Z.R. Grabowski, A. Krowczynski, M. Asher and M. Ottolenghi, *Chem. Phys. Lett.*, 1977, **51**, 315–320.

137. J. Yoon and A.W. Czarnik, *Biorg. Med. Chem.*, 1993, **1**, 267–271.

138. M.W.G. de Bolster, *Pure Appl. Chem.*, 1997, **69**, 1251–1303.

139. Y. Nagai, K. Kobayashi, H. Toi and Y. Aoyama, *Bull. Chem. Soc. Jpn.*, 1993, **66**, 2965–2971.

140. H. Suenaga, M. Mikami, K.R.A.S. Sandanayake and S. Shinkai, *Tetrahedron Lett.*, 1995, **36**, 4825–4828.

141. H. Suenaga, H. Yamamoto and S. Shinkai, *Pure Appl. Chem.*, 1996, **68**, 2179–2186.

142. H. Shinmori, M. Takeuchi and S. Shinkai, *Tetrahedron*, 1995, **51**, 1893–1902.

143. N. DiCesare and J.R. Lakowicz, *J. Phys. Chem. A*, 2001, **105**, 6834–6840.

144. N. DiCesare and J.R. Lakowicz, *J. Photochem. Photobiol., A*, 2001, **143**, 39–47.

145. N. DiCesare and J.R. Lakowicz, *Chem. Commun.*, 2001, 2022–2023.

146. N. DiCesare and J.R. Lakowicz, *Tetrahedron Lett.*, 2002, **43**, 2615–2618.

147. N. DiCesare and J.R. Lakowicz, *Tetrahedron Lett.*, 2001, **42**, 9105–9108.

148. R. Badugu, J.R. Lakowicz and C.D. Geddes, *Anal. Chem.*, 2004, **76**, 610–618.

149. X. Gao, Y. Zhang and B. Wang, *Org. Lett.*, 2003, **5**, 4615–4618.

150. X. Gao, Y. Zhang and B. Wang, *New J. Chem.*, 2005, **29**, 579–586.

151. X. Gao, Y. Zhang and B. Wang, *Tetrahedron*, 2005, **61**, 9111–9117.

152. J. Wang, S. Jin and B. Wang, *Tetrahedron Lett.*, 2005, **46**, 7003–7006.

153. K.R.A.S. Sandanayake, S. Imazu, T.D. James, M. Mikami and S. Shinkai, *Chem. Lett.*, 1995, 139–140.

154. S. Arimori, L.I. Bosch, C.J. Ward and T.D. James, *Tetrahedron Lett.*, 2001, **42**, 4553–4555.

155. L.I. Bosch, M.F. Mahon and T.D. James, *Tetrahedron Lett.*, 2004, **45**, 2859–2862.

156. S. Arimori, L.I. Bosch, C.J. Ward and T.D. James, *Tetrahedron Lett.*, 2002, **43**, 911–913.

157. P.F. Barbara, T.J. Meyer and M.A. Ratner, *J. Phys. Chem.*, 1996, **100**, 13148–13168.

158. S. Fukuzumi, *Org. Biomol. Chem.*, 2003, **1**, 609–620.

159. H. Kurreck and M. Huber, *Angew. Chem., Int. Ed. Engl.*, 1995, **34**, 849–866.

160. M.R. Wasielewski, *Chem. Rev.*, 1992, **92**, 435–461.

161. M.B. Zimmt and D.H. Waldeck, *J. Phys. Chem. A*, 2003, **107**, 3580–3597.

162. H. Oevering, M.N. Paddonrow, M. Heppener, A.M. Oliver, E. Cotsaris, J.W. Verhoeven and N.S. Hush, *J. Am. Chem. Soc.*, 1987, **109**, 3258–3269.

163. G.L. Closs and J.R. Miller, *Science*, 1988, **240**, 440–447.

164. H. Knibbe, D. Rehm and A. Weller, *Ber. Bunsen-Ges. Phys. Chem.*, 1968, **72**, 257–263.

165. A.P. de Silva and R. Rupasinghe, *J. Chem. Soc., Chem. Commun.*, 1985, 1669–1670.

166. J.D. Jackson, *Classical Electrodynamics*, Wiley, New York, 1975.

167. M.C. Beard, G.M. Turner and C.A. Schmuttenmaer, *J. Am. Chem. Soc.*, 2000, **122**, 11541–11542.

168. M.C. Beard, G.M. Turner and C.A. Schmuttenmaer, *J. Phys. Chem. A*, 2002, **106**, 878–883.

169. V. Balzani and F. Scandola, *Supramolecular Photochemistry*, T.J. Kemp (ed), Ellis Horwood Limited, 1991.

170. M.D. Ward, *Chem. Soc. Rev.*, 1997, **26**, 365–376.

171. L. Flamigni, A.M. Talarico, S. Serroni, F. Puntoriero, M.J. Gunter, M.R. Johnston and T.P. Jeynes, *Chem. Eur. J.*, 2003, **9**, 2649–2659.

172. J.W. Verhoeven, *Pure Appl. Chem.*, 1996, **68**, 2223–2286.

173. R.C. Dorfman, Y. Lin and M.D. Fayer, *J. Phys. Chem.*, 1990, **94**, 8007–8009.

174. G. Wulff, *Pure Appl. Chem.*, 1982, **54**, 2093–2102.

175. T.D. James, K. Sandanayake and S. Shinkai, *J. Chem. Soc., Chem. Commun.*, 1994, 477–478.

176. T.D. James, K.R.A.S. Sandanayake, R. Iguchi and S. Shinkai, *J. Am. Chem. Soc.*, 1995, **117**, 8982–8987.

177. H. Kijima, M. Takeuchi, A. Robertson, S. Shinkai, C. Cooper and T.D. James, *Chem. Commun.*, 1999, 2011–2012.

178. J.Z. Zhao, T.M. Fyles and T.D. James, *Angew. Chem., Int. Ed.*, 2004, **43**, 3461–3464.

179. C.W. Gray and T.A. Houston, *J. Org. Chem.*, 2002, **67**, 5426–5428.

180. T.D. James, H. Shinmori and S. Shinkai, *Chem. Commun.*, 1997, 71–72.

181. P. Linnane, T.D. James, S. Imazu and S. Shinaki, *Tetrahedron Lett.*, 1995, **36**, 8833–8834.

182. P. Linnane, T.D. James and S. Shinkai, *J. Chem. Soc., Chem. Commun.*, 1995, 1997–1998.

183. T.D. James, H. Shinmori, M. Takeuchi and S. Shinkai, *Chem. Commun.*, 1996, 705–706.

184. B. Kukrer and E.U. Akkaya, *Tetrahedron Lett.*, 1999, **40**, 9125–9128.

185. R.R. Anderson and J.A. Parrish, *J. Invest. Dermatol.*, 1981, **77**, 13–19.

186. R.M.P. Doornbos, R. Lang, M.C. Aalders, F.W. Cross and H. Sterenborg, *Phys. Med. Biol.*, 1999, **44**, 967–981.

187. J.V. Frangioni, *Curr. Opin. Chem. Biol.*, 2003, **7**, 626–634.

188. J. Zhao, M.G. Davidson, M.F. Mahon, G. Kociok-Köhn and T.D. James, *J. Am. Chem. Soc.*, 2004, **126**, 16179–16186.

189. J.Z. Zhao and T.D. James, *J. Mater. Chem.*, 2005, **15**, 2896–2901.

190. C.R. Cooper and T.D. James, *Chem. Commun.*, 1997, 1419–1420.

191. C.R. Cooper and T.D. James, *J. Chem. Soc., Perkin Trans. 1*, 2000, 963–969.

192. T.D. James and S. Shinkai, *J. Chem. Soc., Chem. Commun.*, 1995, 1483–1485.

193. M. Takeuchi, M. Yamamoto and S. Shinkai, *Chem. Commun.*, 1997, 1731–1732.

194. M. Yamamoto, M. Takeuchi and S. Shinkai, *Tetrahedron*, 1998, **54**, 3125–3140.

195. W. Yang, J. Yan, H. Fang and B. Wang, *Chem. Commun.*, 2003, 792–793.

196. M.J. Deetz and B.D. Smith, *Tetrahedron Lett.*, 1998, **39**, 6841–6844.

197. S.J.M. Koskela, T.M. Fyles and T.D. James, *Chem. Commun.*, 2005, 945–947.

198. C.R. Cooper, N. Spencer and T.D. James, *Chem. Commun.*, 1998, 1365–1366.

199. S. Arimori, M.G. Davidson, T.M. Fyles, T.G. Hibbert, T.D. James and G.I. Kociok-Köhn, *Chem. Commun.*, 2004, 1640–1641.

200. W. Yang, J. Yan, G. Springsteen, S. Deeter and B. Wang, *Bioorg. Med. Chem. Lett.*, 2003, **13**, 1019–1022.

201. W. Yang, L. Lin and B. Wang, *Tetrahedron Lett.*, 2005, **46**, 7981–7984.

202. N. DiCesare, D.P. Adhikari, J.J. Heynekamp, M.D. Heagy and J.R. Lakowicz, *J. Fluorescence*, 2002, **12**, 147–154.

203. D.P. Adhikiri and M.D. Heagy, *Tetrahedron Lett.*, 1999, **40**, 7893–7896.

204. H. Cao, D.I. Diaz, N. DiCesare, J.R. Lakowicz and M.D. Heagy, *Org. Lett.*, 2002, **4**, 1503–1505.

205. H. Cao, T. McGill and M.D. Heagy, *J. Org. Chem.*, 2004, **69**, 2959–2966.

206. A.J. Tong, A. Yamauchi, T. Hayashita, Z.Y. Zhang, B.D. Smith and N. Teramae, *Anal. Chem.*, 2001, **73**, 1530–1536.

207. R. Badugu, J.R. Lakowicz and C.D. Geddes, *J. Fluorescence*, **2003 13**, 371–374.

208. R. Badugu, J.R. Lakowicz and C.D. Geddes, *Talanta*, 2005, **65**, 762–768.

209. R. Badugu, J.R. Lakowicz and C.D. Geddes, *Analyst*, 2004, **129**, 516–521.

210. G. Deng, T.D. James and S. Shinkai, *J. Am. Chem. Soc.*, 1994, **116**, 4567–4572.

211. T.D. James, Y. Shiomi, K. Kondo and S. Shinkai, *XVIII International Symposium on Macrocyclic Chemistry*, University of Twente, Enschede, The Netherlands, 1993.

212. Y. Shiomi, K. Kondo, M. Saisho, T. Harada, K. Tsukagoshi and S. Shinkai, *Supramolecular Chem.*, 1993, **2**, 11–17.

213. Y. Shiomi, M. Saisho, K. Tsukagoshi and S. Shinkai, *J. Chem. Soc., Perkin Trans. 1*, 1993, 2111–2117.

214. K. Sandanayake, K. Nakashima and S. Shinkai, *J. Chem. Soc., Chem. Commun.*, 1994, 1621–1622.

215. N.J. Turro, *Modern Molecular Photochemistry*, Benjamin, Menlo Park, CA, 1978.

216. M. Takeuchi, T. Mizuno, H. Shinmori, M. Nakashima and S. Shinkai, *Tetrahedron*, 1996, **52**, 1195–1204.

217. M. Takeuchi, S. Yoda, T. Imada and S. Shinkai, *Tetrahedron*, 1997, **53**, 8335–8348.

218. A.T. Wright, Z. Zhong and E.V. Anslyn, *Angew. Chem., Int. Ed.*, 2005, **44**, 5679–5682.

219. H. Kijima, M. Takeuchi and S. Shinkai, *Chem. Lett.*, 1998, 781–782.

220. H. Eggert, J. Frederiksen, C. Morin and J.C. Norrild, *J. Org. Chem.*, 1999, **64**, 3846–3852.

221. S. Trupp, A. Schweitzer and G.J. Mohr, *Microchimica Acta*, 2006, **153**, 127–131.

222. E. Nakata, T. Nagase, S. Shinkai and I. Hamachi, *J. Am. Chem. Soc.*, 2004, **126**, 490–495.

223. Z. Wang, D. Zhang and D. Zhu, *J. Org. Chem.*, 2005, **70**, 5729–5732.

224. H. Höpfl, *J. Organomet. Chem.*, 1999, **581**, 129–149.

225. S.L. Wiskur, J.J. Lavigne, H. Ait-Haddou, V. Lynch, Y. Hung Chiu, J.W. Canary and E.V. Anslyn, *Org. Lett.*, 2001, **3**, 1311–1314.

226. S. Franzen, W. Ni and B. Wang, *J. Phys. Chem. B*, 2003, **107**, 12942–12948.

227. W.J. Ni, G. Kaur, G. Springsteen, B.H. Wang and S. Franzen, *Bioorg. Chem.*, 2004, **32**, 571–581.

228. K.L. Bhat, V. Braz, E. Laverty and C.W. Bock, *Theochem-J. Mol. Struct.*, 2004, **712**, 9–19.

229. K.L. Bhat, N.J. Howard, H. Rostami, J.H. Lai and C.W. Bock, *Theochem -J. Mol. Struct.*, 2005, **723**, 147–157.

230. J.D. Morrison and R.L. Letsinger, *J. Org. Chem.*, 1964, **29**, 3405–3407.

231. L. Zhu, S.H. Shabbir, M. Gray, V.M. Lynch, S. Sorey and E.V. Anslyn, *J. Am. Chem. Soc.*, 2006, **128**, 1222–1232.

232. M. Bielecki, H. Eggert and J.C. Norrild, *J. Chem. Soc., Perkin Trans. 2*, 1999, 449–455.

233. J.C. Norrild and H. Eggert, *J. Am. Chem. Soc.*, 1995, **117**, 1479–1484.

234. M.P. Nicholls and P.K.C. Paul, *Org. Biomol. Chem.*, 2004, **2**, 1434–1441.

235. S.J. Angyal, *Adv. Carbohydr. Chem. Biochem.*, 1984, **42**, 15–68.

236. J.C. Norrild and H. Eggert, *J. Chem. Soc., Perkin Trans. 2*, 1996, 2583–2588.

237. C.R. Cooper and T.D. James, *Chem. Lett.*, 1998, 883–884.

238. W. Yang, H. He and D.G. Drueckhammer, *Angew. Chem., Int. Ed.*, 2001, **40**, 1714–1718.

239. K.R.A.S. Sandanayake, T.D. James and S. Shinkai, *Chem. Lett.*, 1995, 503–504.

240. B. Appleton and T.D. Gibson, *Sens. Actuators, B*, 2000, **65**, 302–304.

241. M.D. Phillips, Fluorescent phosphate sensors, M.NatSc(Chem) Dissertation, The University of Birmingham, Birmingham, 2000.

242. S. Arimori, M.L. Bell, C.S. Oh, K.A. Frimat and T.D. James, *Chem. Commun.*, 2001, 1836–1837.

243. S. Arimori, M.L. Bell, C.S. Oh, K.A. Frimat and T.D. James, *J. Chem. Soc., Perkin Trans. 1*, 2002, 803–808.
244. D.D. Perrin and B. Dempsey, *Buffers for Ph and Metal Ion Control*, Chapman & Hall, 1974.
245. G. Lecollinet, A.P. Dominey, T. Velasco and A.P. Davis, *Angew. Chem., Int. Ed.*, 2002, **41**, 4093–4096.
246. Z. Zhong and E.V. Anslyn, *J. Am. Chem. Soc.*, 2002, **124**, 9014–9015.
247. A. Sugasaki, K. Sugiyasu, M. Ikeda, M. Takeuchi and S. Shinkai, *J. Am. Chem. Soc.*, 2001, **123**, 10239–10244.
248. A. Sugasaki, M. Ikeda, M. Takeuchi and S. Shinkai, *Angew. Chem., Int. Ed.*, 2000, **39**, 3839–3842.
249. M. Yamamoto, M. Takeuchi and S. Shinkai, *Tetrahedron*, 2002, **58**, 7251–7258.
250. M. Ikeda, S. Shinkai and A. Osuka, *Chem. Commun.*, 2000, 1047–1048.
251. W. Yang, S. Gao, X. Gao, V.V.R. Karnati, W. Ni, B. Wang, W.B. Hooks, J. Carson and B. Weston, *Bioorg. Med. Chem. Lett.*, 2002, **12**, 2175–2177.
252. S. Patterson, B.D. Smith and R.E. Taylor, *Tetrahedron Lett.*, 1998, **39**, 3111–3114.
253. T. Nagase, E. Nakata, S. Shinkai and I. Hamachi, *Chem. Eur. J.*, 2003, **9**, 3660–3669.
254. S. Arimori, M.D. Phillips and T.D. James, *Tetrahedron Lett.*, 2004, **45**, 1539–1542.
255. S. Arimori, M.L. Bell, C.S. Oh and T.D. James, *Org. Lett.*, 2002, **4**, 4249–4251.
256. T. Jin, *Chem. Commun.*, 1999, 2491–2492.
257. S. Arimori, G.A. Consiglio, M.D. Phillips and T.D. James, *Tetrahedron Lett.*, 2003, **44**, 4789–4792.
258. M. Barboiu, C.T. Supuran, A. Scozzafava, F. Briganti, C. Luca, G. Popescu, L. Cot and N. Hovnanian, *Liebigs Ann. Chem.*, 1997, 1853–1859.
259. S.A. Galema, M.J. Blandamer and J. Engberts, *J. Am. Chem. Soc.*, 1990, **112**, 9665–9666.
260. D. Fabri, M.A.K. Williams and T.K. Halstead, *Carbohydr. Res.*, 2005, **340**, 889–905.
261. S.A. Galema, M.J. Blandamer and J. Engberts, *J. Org. Chem.*, 1992, **57**, 1995–2001.
262. M. Janado and Y. Yano, *Bull. Chem. Soc. Jpn.*, 1985, **58**, 1913–1917.
263. Y. Yano, K. Tanaka, Y. Doi and M. Janado, *Bull. Chem. Soc. Jpn.*, 1988, **61**, 2963–2964.
264. E. Fischer, *Ber. Dtsch. Chem. Ges.*, 1894, **27**, 2985–2993.
265. R.U. Lemieux and U. Spohr, *Adv. Carbohydr. Chem. Biochem.*, 1994, **50**, 1–20.
266. D.T. Warner, *Nature*, 1962, **196**, 1055–1058.
267. M.D. Walkinshaw, *J. Chem. Soc., Perkin Trans. 2*, 1987, 1903–1905.
268. R.U. Lemieux, *Acc. Chem. Res.*, 1996, **29**, 373–380.

269. P.R. Westmark and B.D. Smith, *J. Am. Chem. Soc.*, 1994, **116**, 9343–9344.

270. S.P. Draffin, P.J. Duggan and S.A.M. Duggan, *Org. Lett.*, 2001, **3**, 917–920.

271. S.J. Gardiner, B.D. Smith, P.J. Duggan, M.J. Karpa and G.J. Griffin, *Tetrahedron*, 1999, **55**, 2857–2864.

272. P.R. Westmark, S.J. Gardiner and B.D. Smith, *J. Am. Chem. Soc.*, 1996, **118**, 11093–11100.

273. J.A. Riggs, R.K. Litchfield and B.D. Smith, *J. Org. Chem.*, 1996, **61**, 1148–1150.

274. G.T. Morin, M.P. Hughes, M.F. Paugam and B.D. Smith, *J. Am. Chem. Soc.*, 1994, **116**, 8895–8901.

275. V.V. Karnati, X. Gao, S. Gao, W. Yang, W. Ni, S. Sankar and B. Wang, *Bioorg. Med. Chem. Lett.*, 2002, **12**, 3373–3377.

276. W. Yang, H. Fan, X. Gao, S. Gao, V.V.R. Karnati, W. Ni, W.B. Hooks, J. Carson, B. Weston and B. Wang, *Chem. Biol.*, 2004, **11**, 439–448.

277. D. Stones, S. Manku, X.S. Lu and D.G. Hall, *Chem. Eur. J.*, 2004, **10**, 92–100.

278. F.A. Carey and R.J. Sundberg, *Advanced Organic Chemistry*, Kluwer Academic/Plenum Publishers, 2000.

279. H.S. Snyder and S.L. Meisel, *J. Am. Chem. Soc.*, 1948, **70**, 774–776.

280. H.R. Snyder and C. Weaver, *J. Am. Chem. Soc.*, 1948, **70**, 232–234.

281. A.P. Russell (ed), WO 91/04488, 1991.

282. T. Nagasaki, H. Shinmori and S. Shinkai, *Tetrahedron Lett.*, 1994, **9**, 2201–2204.

283. M. Takeuchi, M. Taguchi, H. Shinmori and S. Shinkai, *Bull. Chem. Soc. Jpn.*, 1996, **69**, 2613–2618.

284. K.R.A.S. Sandanayake and S. Shinkai, *J. Chem. Soc., Chem. Commun.*, 1994, 1083–1084.

285. H. Shinmori, M. Takeuchi and S. Shinkai, *J. Chem. Soc., Perkin Trans. 2*, 1998, 847–852.

286. K. Koumoto and S. Shinkai, *Chem. Lett.*, 2000, 856–857.

287. K. Koumoto, M. Takeuchi and S. Shinkai, *Supramolecular Chem.*, 1998, **9**, 203 (210pp).

288. C.J. Ward, P. Patel, P.R. Ashton and T.D. James, *Chem. Commun.*, 2000, 229–230.

289. C.J. Ward, P. Patel and T.D. James, *J. Chem. Soc., Perkin Trans. 1*, 2002, 462–470.

290. N. DiCesare and J.R. Lakowicz, *Org. Lett.*, 2001, **3**, 3891–3893.

291. H. Shinmori, M. Takeuchi and S. Shinkai, *J. Chem. Soc., Perkin Trans. 2*, 1996, 1–3.

292. H. Yamamoto, A. Ori, K. Ueda, C. Dusemund and S. Shinkai, *Chem. Commun.*, 1996, 407–408.

293. T. Mizuno, M. Takeuchi and S. Shinkai, *Tetrahedron*, 1999, **55**, 9455–9468.

294. T. Mizuno, M. Yamamoto, M. Takeuchi and S. Shinkai, *Tetrahedron*, 2000, **56**, 6193–6198.

295. V.W.W. Yam and A.S.F. Kai, *Chem. Commun.*, 1998, 109–110.

296. T. Mizuno, T. Fukumatsu, M. Takeuchi and S. Shinkai, *J. Chem. Soc., Perkin Trans. 1*, 2000, 407–413.

297. C.J. Davis, P.T. Lewis, M.E. McCarroll, M.W. Read, R. Cueto and R.M. Strongin, *Org. Lett.*, 1999, **1**, 331–334.

298. O. Rusin, O. Alpturk, M. He, J.O. Escobedo, S. Jiang, F. Dawan, K. Lian, M.E. McCarroll, I.M. Warner and R.M. Strongin, *J. Fluorescence*, 2004, **14**, 611–615.

299. P.T. Lewis, C.J. Davis, L.A. Cabell, M. He, M.W. Read, M.E. McCarroll and R.M. Strongin, *Org. Lett.*, 2000, **2**, 589–592.

300. M. He, J.R. Johnson, J.O. Escobedo, P.A. Beck, K.K. Kim, N.N. St. Luce, C.J. Davis, P.T. Lewis, F.R. Fronczeck, B.J. Melancon, A.A. Mrse, W.D. Treleaven and R.M. Strongin, *J. Am. Chem. Soc.*, 2002, **124**, 5000–5009.

301. C.J. Ward, P. Patel and T.D. James, *Org. Lett.*, 2002, **4**, 477–479.

302. K. Sato, A. Sone, S. Arai and T. Yamagishi, *Heterocycles*, 2003, **61**, 31–38.

303. W. Ni, H. Fang, G. Springsteen and B. Wang, *J. Org. Chem.*, 2004, **69**, 1999–2007.

304. G.S. Wilson and H. Yibai, *Chem. Rev.*, 2000, **100**, 2693–2704.

305. A. Ori and S. Shinkai, *J. Chem. Soc., Chem. Commun.*, 1995, 1771–1772.

306. A.N.J. Moore and D.D.M. Wayner, *Can. J. Chem.*, 1999, **77**, 681–686.

307. M. Nicolas, B. Fabre and J. Simonet, *Electrochim. Acta*, 2001, **46**, 1179–1190.

308. S. Arimori, S. Ushiroda, L.M. Peter, A.T.A. Jenkins and T.D. James, *Chem. Commun.*, 2002, 2368–2369.

309. J.C. Norrild and I. Sotofte, *J. Chem. Soc., Perkin Trans. 2*, 2002, 303–311.

310. J.J. Lavigne and E.V. Anslyn, *Angew. Chem., Int. Ed.*, 2001, **40**, 3118–3130.

311. S.L. Wiskur, H. Ait-Haddou, J.J. Lavigne and E.V. Anslyn, *Acc. Chem. Res.*, 2001, **34**, 963–972.

312. A.T. Wright and E.V. Anslyn, *Chem. Soc. Rev.*, 2006, **35**, 14–28.

313. S.L. Wiskur, J.J. Lavigne, A. Metzger, S.L. Tobey, V. Lynch and E.V. Anslyn, *Chem. Eur. J.*, 2004, **10**, 3792–3804.

314. L.A. Cabell, M.K. Monahan and E.V. Anslyn, *Tetrahedron Lett.*, 1999, **40**, 7753–7756.

315. J.J. Lavigne and E.V. Anslyn, *Angew. Chem. Int. Ed.*, 1999, **38**, 3666–3669.

316. A.M. Piatek, Y.J. Bomble, S.L. Wiskur and E.V. Anslyn, *J. Am. Chem. Soc.*, 2004, **126**, 6072–6077.

317. S.L. Wiskur and E.V. Anslyn, *J. Am. Chem. Soc.*, 2001, **123**, 10109–10110.

318. S.L. Wiskur, P.N. Floriano, E.V. Anslyn and J.T. McDevitt, *Angew. Chem., Int. Ed.*, 2003, **42**, 2070–2072.

319. B.T. Nguyen, S.L. Wiskur and E.V. Anslyn, *Org. Lett.*, 2004, **6**, 2499–2501.

320. L. Zhu and E.V. Anslyn, *J. Am. Chem. Soc.*, 2004, **126**, 3676–3677.
321. L. Zhu, Z. Zhong and E.V. Anslyn, *J. Am. Chem. Soc.*, 2005, **127**, 4260–4269.
322. S. Arimori, H. Murakami, M. Takeuchi and S. Shinkai, *J. Chem. Soc., Chem. Commun.*, 1995, 961–962.
323. Z. Murtaza, L. Tolosa, P. Harms and J. R. Lakowicz, *J. Fluorescence*, 2002, **12**, 187–192.
324. J.T. Suri, D.B. Cordes, F.E. Cappuccio, R.A. Wessling and B. Singaram, *Langmuir*, 2003, **19**, 5145–5152.
325. J.N. Camara, J.T. Suri, F.E. Cappuccio, R.A. Wessling and B. Singaram, *Tetrahedron Lett.*, 2002, **43**, 1139–1141.
326. F.E. Cappuccio, J.T. Suri, D.B. Cordes, R.A. Wessling and B. Singaram, *J. Fluorescence*, 2004, **14**, 521–533.
327. D.B. Cordes, S. Gamsey, Z. Sharrett, A. Miller, P. Thoniyot, R.A. Wessling and B. Singaram, *Langmuir*, 2005, **21**, 6540–6547.
328. D.B. Cordes, A. Miller, S. Gamsey, Z. Sharrett, P. Thoniyot, R. Wessling and B. Singaram, *Org. Biomol. Chem.*, 2005, **3**, 1708–1713.
329. N. DiCesare, M.R. Pinto, K.S. Schanze and J.R. Lakowicz, *Langmuir*, 2002, **18**, 7785–7787.
330. G. Springsteen and B. Wang, *Chem. Commun.*, 2001, 1608–1609.
331. J. Yan, G. Springsteen, S. Deeter and B. Wang, *Tetrahedron*, 2004, **60**, 11205–11209.
332. S. Arimori, C.J. Ward and T.D. James, *Chem. Commun.*, 2001, 2018–2019.
333. S. Boduroglu, J.M. El Khoury, D. Venkat Reddy, P.L. Rinaldi and J. Hu, *Bioorg. Med. Chem. Lett.*, 2005, **15**, 3974–3977.
334. S. Arimori, C.J. Ward and T.D. James, *Tetrahedron Lett.*, 2002, **43**, 303–305.
335. H.R. Mulla, N.J. Agard and A. Basu, *Bioorg. Med. Chem. Lett.*, 2004, **14**, 25–27.
336. S. Patterson, B.D. Smith and R.E. Taylor, *Tetrahedron Lett.*, 1997, **38**, 6323–6326.
337. T. Nagasaki, T. Kimura, S. Arimori and S. Shinkai, *Chem. Lett.*, 1994, 1495–1498.
338. T. Kimura, S. Arimori, M. Takeuchi, T. Nagasaki and S. Shinkai, *J. Chem. Soc., Perkin Trans. 2*, 1995, 1889–1894.
339. T. Kimura, M. Takeuchi, T. Nagasaki and S. Shinkai, *Tetrahedron Lett.*, 1995, **36**, 559–562.
340. W. Wang, S. Gao and B. Wang, *Org. Lett.*, 1999, **1**, 1209–1212.
341. S. Gao, W. Wang and B. Wang, *Bioorg. Chem.*, 2001, **29**, 308–320.
342. T. Kawanishi, M.A. Romey, P.C. Zhu, M.Z. Holody and S. Shinkai, *J. Fluorescence*, 2004, **14**, 499–512.
343. A.T. Wright, M.J. Griffin, Z. Zhong, S.C. McCleskey, E.V. Anslyn and J.T. McDevitt, *Angew. Chem., Int. Ed.*, 2005, **44**, 6375–6378.
344. J.T. Suri, D.B. Cordes, F.E. Cappuccio, R.A. Wessling and B. Singaram, *Angew. Chem., Int. Ed.*, 2003, **42**, 5857–5859.

345. S.A. Asher, V.L. Alexeev, A.V. Goponenko, A.C. Sharma, I.K. Lednev, C.S. Wilcox and D.N. Finegold, *J. Am. Chem. Soc.*, 2003, **125**, 3322–3329.
346. V.L. Alexeev, A.C. Sharma, A.V. Goponenko, S. Das, I.K. Lednev, C.S. Wilcox, D.N. Finegold and S.A. Asher, *Anal. Chem.*, 2003, **75**, 2316–2323.
347. V.L. Alexeev, S. Das, D.N. Finegold and S.A. Asher, *Clin. Chem. (Washington, DC, United States)*, 2004, **50**, 2353–2360.
348. E. Pringsheim, E. Terpetschnig, S.A. Piletsky and O.S. Wolfbeis, *Adv. Mater.*, 1999, **11**, 865–868.
349. J. Zhang, C.D. Geddes and J.R. Lakowicz, *Anal. Biochem.*, 2004, **332**, 253–260.
350. K. Aslan, J. Zhang, J.R. Lakowicz and C.D. Geddes, *J. Fluorescence*, 2004, **14**, 391–400.
351. H. Murakami, H. Akiyoshi, T. Wakamatsu, T. Sagara and N. Nakashima, *Chem. Lett.*, 2000, 940–941.
352. E. Shoji and M.S. Freund, *J. Am. Chem. Soc.*, 2001, **123**, 3383–3384.
353. E. Shoji and M.S. Freund, *J. Am. Chem. Soc.*, 2002, **124**, 12486–12493.
354. B. Fabre and L. Taillebois, *Chem. Commun.*, 2003, 2982–2983.
355. A. Kikuchi, K. Suzuki, O. Okabayashi, H. Hoshino, K. Kataoka, Y. Sakurai and T. Okano, *Anal. Chem.*, 1996, **68**, 823–828.
356. Y. Ma and X. Yang, *J. Electroanal. Chem.*, 2005, **580**, 348–352.
357. F.H. Arnold, W. Zheng and A.S. Michaels, *J. Memb. Sci.*, 2000, **167**, 227–239.
358. R. Gabai, N. Sallacan, V. Chegel, T. Bourenko, E. Katz and I. Willner, *J. Phys. Chem. B*, 2001, **105**, 8196–8202.
359. M. Lee, T.-I. Kim, K.-H. Kim, J.-H. Kim, M.-S. Choi, H.-J. Choi and K. Koh, *Anal. Biochem.*, 2002, **310**, 163–170.
360. Y. Ma, N. Li, C. Yang and X. Yang, *Colloids and Surfaces A: Physicochemical and Engineering Aspects*, 2005, **269**, 1–6.
361. N. Sallacan, M. Zayats, T. Bourenko, A.B. Kharitonov and I. Willner, *Anal. Chem.*, 2002, **74**, 702–712.
362. J. Pribyl and P. Skladal, *Anal. Chim. Acta*, 2005, **530**, 75–84.
363. M.-C. Lee, S. Kabilan, A. Hussain, X. Yang, J. Blyth and C.R. Lowe, *Anal. Chem.*, 2004, **76**, 5748–5755.
364. M. Zayats, M. Lahav, A.B. Kharitonov and I. Willner, *Tetrahedron*, 2002, **58**, 815–824.
365. S. Zhang, R. Trokowski and A.D. Sherry, *J. Am. Chem. Soc.*, 2003, **125**, 15288–15289.
366. L. Frullano, J. Rohovec, S. Aime, T. Maschmeyer, M.I. Prata, J.J. Pedroso de Lima, C.F.G.C. Geraldes and J.A. Peters, *Chem. Eur. J.*, 2004, **10**, 5205–5217.
367. Y. Perez-Fuertes, A.M. Kelly, A.L. Johnson, S. Arimori, S.D. Bull and T.D. James, *Org. Lett.*, 2006, **8**, 609–612.
368. C. Cannizzo, S. Amigoni-Gerbier and C. Larpent, *Polymer*, 2005, **46**, 1269–1276.
369. Y. Ma, Q. Gao and X. Yang, *Microchimica Acta*, 2005, **150**, 21–26.

370. K. Kataoka, H. Miyazaki, M. Bunya, T. Okana and Y. Sakurai, *J. Am. Chem. Soc.*, 1998, **120**, 12694–12695.

371. A. Matsumoto, S. Ikeda, A. Harada and K. Kataoka, *Biomacromolecules*, 2003, **4**, 1410–1416.

372. A. Laschewsky, *Angew. Chem. Int. Ed. Engl.*, 1989, **28**, 1574–1577.

373. R. Ludwig, T. Harada, K. Ueda, T.D. James and S. Shinkai, *J. Chem. Soc., Perkin Trans. 2*, 1994, 697–702.

374. R. Ludwig, K. Ariga and S. Shinkai, *Chem. Lett.*, 1993, 1413–1416.

375. R. Ludwig, Y. Shiomi and S. Shinkai, *Langmuir*, 1994, **10**, 3195–3200.

376. C. Dusemund, M. Mikami and S. Shinkai, *Chem. Lett.*, 1995, 157–158.

377. T. Miyahara and K. Kurihara, *J. Am. Chem. Soc.*, 2004, **126**, 5684–5685.

378. M. Pietraszkiewicz, P. Prus and O. Pietraszkiewicz, *Tetrahedron*, 2004, **60**, 10747–10752.

379. T. Araki and H. Tsukube, *Liquid Membranes Chemical Applications*, CRC press, Boca Raton, FL, 1990.

380. T. Shinbo, K. Nishimura, T. Yamaguchi and M. Sugiura, *J. Chem. Soc., Chem. Commun.*, 1986, 349–351.

381. B.F. Grotjohn and A.W. Czarnik, *Tetrahedron Lett.*, 1989, **30**, 2325–2328.

382. M.F. Paugam, L.S. Valencia, B. Boggess and B.D. Smith, *J. Am. Chem. Soc.*, 1994, **116**, 11203–11204.

383. M.F. Paugam and B.D. Smith, *Tetrahedron Lett.*, 1993, **34**, 3723–3726.

384. L.K. Mohler and A.W. Czarnik, *J. Am. Chem. Soc.*, 1993, **115**, 2998–2999.

385. L.K. Mohler and A.W. Czarnik, *J. Am. Chem. Soc.*, 1993, **115**, 7037–7038.

386. M.F. Paugam, J.A. Riggs and B.D. Smith, *Chem. Commun.*, 1996, 2539–2540.

387. J.A. Riggs, K.A. Hossler, B.D. Smith, M.J. Karpa, G. Griffin and P.J. Duggan, *Tetrahedron Lett.*, 1996, **37**, 6303–6306.

388. M. Takeuchi, K. Koumoto, M. Goto and S. Shinkai, *Tetrahedron*, 1996, **52**, 12931–12940.

389. J.T. Bien, M.Y. Shang and B.D. Smith, *J. Org. Chem.*, 1995, **60**, 2147–2152.

390. M.F. Paugam, J.T. Bien, B.D. Smith, L.A.J. Chrisstoffels, F. de Jong and D.N. Reinhoudt, *J. Am. Chem. Soc.*, 1996, **118**, 9820–9825.

391. M. Di Luccio, B.D. Smith, T. Kida, C.P. Borges and T.L.M. Alves, *J. Membr. Sci.*, 2000, **174**, 217–224.

392. S.P. Draffin, P.J. Duggan, S.A.M. Duggan and J.C. Norrild, *Tetrahedron*, 2003, **59**, 9075–9082.

393. P.J. Duggan, *Aust. J. Chem.*, 2004, **57**, 291–299.

394. B.D. Smith, J.P. Davis, S.P. Draffin and P.J. Duggan, *Supramolecular Chem.*, 2004, **16**, 87–90.

395. T.M. Altamore, E.S. Barrett, P.J. Duggan, M.S. Sherburn and M.L. Szydzik, *Org. Lett.*, 2002, **4**, 3489–3491.

396. K. Nakashima and S. Shinkai, *Chem. Lett.*, 1995, 443–444.

397. K. Nakashima and S. Shinkai, *Chem. Lett.*, 1994, 1267–1270.

398. T. Mizuno, M. Takeuchi, I. Hamachi, K. Nakashima and S. Shinkai, *Chem. Commun.*, 1997, 1793–1794.
399. T. Mizuno, M. Takeuchi, I. Hamachi, K. Nakashima and S. Shinkai, *J. Chem. Soc., Perkin Trans. 2*, 1998, 2281–2288.
400. G. Nuding, K. Nakashima, R. Iguchi, T. Ishi-i and S. Shinkai, *Tetrahedron Lett.*, 1998, **39**, 9473–9476.
401. M. Yamamoto, M. Takeuchi and S. Shinkai, *Tetrahedron Lett.*, 1998, **39**, 1189–1192.
402. M. Yamamoto, M. Takeuchi, S. Shinkai, F. Tani and Y. Naruta, *J. Chem. Soc., Perkin Trans. 2*, 2000, 9–16.
403. F. Ohseto, H. Yamamoto, H. Matsumoto and S. Shinkai, *Tetrahedron Lett.*, 1995, **36**, 6911–6914.
404. T. Imada, H. Kijima, M. Takeuchi and S. Shinkai, *Tetrahedron Lett.*, 1995, **36**, 2093–2096.
405. T. Imada, H. Kijima, M. Takeuchi and S. Shinkai, *Tetrahedron*, 1996, **52**, 2817–2826.
406. M. Takeuchi, H. Kijima, I. Hamachi and S. Shinkai, *Bull. Chem. Soc. Jpn.*, 1997, **70**, 699–705.
407. L.D. Sarson, K. Ueda, R. Takeuchi and S. Shinkai, *Chem. Commun.*, 1996, 619–620.
408. S. Arimori, M. Takeuchi and S. Shinkai, *Chem. Lett.*, 1996, 77–78.
409. H. Suenaga, S. Arimori and S. Shinkai, *J. Chem. Soc., Perkin Trans. 2*, 1996, 607–612.
410. M. Takeuchi, Y. Chin, T. Imada and S. Shinkai, *Chem. Commun.*, 1996, 1867–1868.
411. M. Takeuchi, S. Yoda, Y. Chin and S. Shinkai, *Tetrahedron Lett.*, 1999, **40**, 3745–3748.
412. M. Takeuchi, T. Imada and S. Shinkai, *J. Am. Chem. Soc.*, 1996, **118**, 10658–10659.
413. M. Takeuchi, T. Imada and S. Shinkai, *Bull. Chem. Soc. Jpn.*, 1998, **71**, 1117–1123.
414. O. Hirata, Y. Kubo, M. Takeuchi and S. Shinkai, *Tetrahedron*, 2004, **60**, 11211–11218.
415. R.G. Weiss, *Tetrahedron*, 1988, **44**, 3413–3475.
416. T.D. James, H. Kawabata, R. Ludwig, K. Murata and S. Shinkai, *Tetrahedron*, 1995, **51**, 555–566.
417. T.D. James, K. Murata, T. Harada, K. Ueda and S. Shinkai, *Chem. Lett.*, 1994, 273–276.
418. T. Imada, H. Murakami and S. Shinkai, *J. Chem. Soc., Chem. Commun.*, 1994, 1557–1558.
419. K. Koumoto, T. Yamashita, T. Kimura, R. Luboradzki and S. Shinkai, *Nanotechnology*, 2001, **12**, 25–31.
420. T. Kimura and S. Shinkai, *Chem. Lett.*, 1998, 1035–1036.
421. T. Kimura, T. Yamashita, K. Koumoto and S. Shinkai, *Tetrahedron Lett.*, 1999, **40**, 6631–6634.

422. T. Kimura, M. Takeuchi and S. Shinkai, *Bull. Chem. Soc. Jpn.*, 1998, **71**, 2197–2204.

423. S. Arimori, M. Takeuchi and S. Shinkai, *J. Am. Chem. Soc.*, 1996, **118**, 245–246.

424. S. Arimori, M. Takeuchi and S. Shinkai, *Supramolecular Sci.*, 1998, **5**, 1–8.

425. T.D. James, T. Harada and S. Shinkai, *J. Chem. Soc., Chem. Commun.*, 1993, 857–860.

426. H. Kobayashi, M. Amaike, J.H. Jung, A. Friggeri, S. Shinkai and D.N. Reinhoudt, *Chem. Commun.*, 2001, 1038–1039.

427. H. Kobayashi, K. Koumoto, J. H. Jung and S. Shinkai, *J. Chem. Soc., Perkin Trans. 2*, 2002, 1930–1936.

428. G. Wulff, W. Vesper, R. Grobeeinsler and A. Sarhan, *Macromol. Chem. Physic.*, 1977, **178**, 2799–2816.

429. G. Wulff, *Angew. Chem., Int. Ed. Engl.*, 1995, **34**, 1812–1832.

430. G. Wulff, *Chem. Rev.*, 2002, **102**, 1–28.

431. T. Ishi-i, K. Nakashima, S. Shinkai and A. Ikeda, *J. Org. Chem.*, 1999, **64**, 984–990.

432. T. Ishi-i, R. Iguchi and S. Shinkai, *Tetrahedron*, 1999, **55**, 3883–3892.

433. Y. Kanekiyo, Y. Ono, K. Inoue, M. Sano and S. Shinkai, *J. Chem. Soc., Perkin Trans. 2*, 1999, 557–561.

434. Y. Kanekiyo, M. Sano, R. Iguchi and S. Shinkai, *J. Polym. Sci. Pol. Chem.*, 2000, **38**, 1302–1310.

435. A. Friggeri, H. Kobayashi, S. Shinkai and D.N. Reinhoudt, *Angew. Chem., Int. Ed.*, 2001, **40**, 4729–4731.

436. H. Kobayashi, K. Nakashima, E. Ohshima, Y. Hisaeda, I. Hamachi and S. Shinkai, *J. Chem. Soc., Perkin Trans. 2*, 2000, **5**, 997–1002.

437. K. Dabulis and A.M. Klibanov, *Biotechnol. Bioeng.*, 1992, **39**, 176–185.

438. H.Y. Wang, T. Kobayashi, T. Fukaya and N. Fujii, *Langmuir*, 1997, **13**, 5396–5400.

439. Y. Kanekiyo, M. Sano, Y. Ono, K. Inoue and S. Shinkai, *J. Chem. Soc., Perkin Trans. 2*, 1998, 2005–2008.

440. K. Inoue, Y. Ono, Y. Kanekiyo, T. IshiI, K. Yoshihara and S. Shinkai, *Tetrahedron Lett.*, 1998, **39**, 2981–2984.

Subject Index

Author Biographies

Tony D. James was born in Broseley, Shropshire England in 1964. He received his BSc in 1986 from the University of East Anglia and PhD in 1991 from the University of Victoria, Canada under the supervision of Thomas M. Fyles. After postdoctoral work with Seiji Shinkai at his Chemirecognics Project in Kurume, Japan he returned to the University of Birmingham as a Royal Society Researcher in 1995. He then moved to the University of Bath in 2000 to take a Lectureship in Organic Chemistry and was subsequently promoted to Senior Lecturer in 2005. He has published four book chapters and over 70 papers in international peer reviewed journals. Citation statistics (August 2006) indicate that two of his publications have been cited over 200 times, seven over 100, and 16 over 50, with a total of more than 2500 citations from 74 papers at a frequency of 34.2 citations per paper. His current h-index is 26. Research interests include: supramolecular chemistry, sensor design, boronic acid-based synthetic receptors, chiral recognition, saccharide recognition, anion recognition, and asymmetric catalysis.

Marcus D. Phillips was born in 1978 in London, England. The son of an English father and a Spanish mother he was raised in Bridgnorth, Shropshire. He obtained his MNatSc(Chemistry) from The University of Birmingham in 2000 before taking a gap year to work and travel through South East Asia, Australasia and South America. On returning to the UK he undertook his PhD in Organic Chemistry at the University of Bath under the supervision of Tony D. James. Research at the University of Bath and at Saitama University, Japan explored the synthetic methodologies required to construct boronic acid-based fluorescent sensors and the host–guest interactions between these sensors and specific saccharides. Following the award of his PhD in 2005 he continued research into the molecular recognition of saccharides with a Postdoctoral Fellowship at the University of Bath. In September 2005 he was appointed to his current position of Research and Development Chemist with Clariant UK Ltd at the Company's manufacturing site and centre of excellence for phenol chemistry in Selby, North Yorkshire.

Seiji Shinkai was born in 1944 in Fukuoka, Japan, and received his PhD in 1972 from Kyushu University, where he became a lecturer soon afterwards. After postdoctoral work at the University of California, Santa Barbara, with Thomas C. Bruice, he joined Kyushu University in 1975 and became a full professor there in 1988. He also worked at the director of Shinkai Chemi-recognics Project (a government-owned ERATO Project under the Japanese Research Development Corporation (JRDC)) (1990–1995). Subsequently, he had served as the director of Chemotransfiguration Project (Japan Science and Technology (JST) Corporation, which was reorganized from JST) (1997–2001), which was an International Collaboration Project (ICORP) with Twente University (director: Professor David N. Reinhoudt). As an extension program of ICORP, he has launched a SORST Project related to sugar-based gene manipulators from March 2002. He had also been acting as a leader of the Kyushu University COE Project entitled "Design and Control of Advanced Molecular Assembly Systems" (1998–2002). From 2002 he is serving as a leader of the 21st Century COE Program entitled "Functional Innovation of Molec-ular Informatics". He has published 875 original papers and 163 reviews and books by the end of 2005. His research interests focus on host–guest chemistry, molecular recognition, sugar sensing, allosteric functions, organogels, sol–gel transcription, and polysaccharide–polynucleotide interactions. He is particu-larly well-known as the first "designer" of a molecular machine system featur-ing photoresponsive crown ethers.